TECHNICAL
REPORT

Who Is "Joint"?

New Evidence from the 2005 Joint Officer Management Census Survey

Sheila Nataraj Kirby, Al Crego, Harry J. Thie,
Margaret C. Harrell, Kimberly Curry,
Michael S. Tseng

Prepared for the Office of the Secretary of Defense

RAND NATIONAL DEFENSE RESEARCH INSTITUTE

The research described in this report was prepared for the Office of the Secretary of Defense (OSD). The research was conducted in the RAND National Defense Research Institute, a federally funded research and development center sponsored by the Office of the Secretary of Defense, the Joint Staff, the Unified Combatant Commands, the Department of the Navy, the Marine Corps, the defense agencies, and the defense Intelligence Community under Contract DASW01-01-C-0004.

Library of Congress Cataloging-in-Publication Data is available for this publication.

ISBN 0-8330-3919-9

Published 2006 by the RAND Corporation
1776 Main Street, P.O. Box 2138, Santa Monica, CA 90407-2138
1200 South Hayes Street, Arlington, VA 22202-5050
4570 Fifth Avenue, Suite 600, Pittsburgh, PA 15213-1516
RAND URL: http://www.rand.org/
To order RAND documents or to obtain additional information, contact
Distribution Services: Telephone: (310) 451-7002;
Fax: (310) 451-6915; Email: order@rand.org

Preface

Several recent studies, including a study authorized under the 2002 National Defense Authorization Act, have indicated the need for the Department of Defense (DoD) to update the practice, policy, and law applied to Joint Officer Management (JOM) and Joint Professional Military Education (JPME) to meet the demands of a new era more effectively.

In 2003, DoD asked the RAND National Defense Research Institute to undertake an analysis that would provide overarching guidance on officer training and development in joint matters. The results of that effort were documented in *Framing a Strategic Approach for Joint Officer Management* (Thie et al., 2005). This work builds on that earlier effort.

As a lead-in to this effort, in summer 2005, the research sponsor and another organization conducted the Joint Officer Management (JOM) Census survey of individuals serving in billets that were likely to either require prior joint experience or provide officers with joint experience. This report provides an overview of the survey responses, including the extent to which officers believe that their assignments provide them with joint experience or require them to have prior joint education, training, or experience. As such, this report should be of interest particularly to military personnel managers dealing with joint officer management issues.

Follow-up work by the RAND Corporation will examine the extent to which prior "jointness" is required by billets and whether sufficient numbers of officers with joint education, training, and experience are likely to be available to satisfy DoD's needs.

This research was sponsored by the Undersecretary of Defense for Personnel and Readiness. It was conducted within the Forces and Resources Policy Center of the RAND National Defense Research Institute, a federally funded research and development center sponsored by the Office of the Secretary of Defense, the Joint Staff, the Unified Combatant Commands, the Department of the Navy, the Marine Corps, the defense agencies, and the defense Intelligence Community. The principal investigators are Harry Thie and Margaret Harrell. Comments are welcome and may be addressed to Harry Thie at harry_thie@rand.org, to Margaret Harrell at margaret_harrell@rand.org, or to the lead author, Sheila Kirby, at sheila_kirby@rand.org.

For more information on RAND's Forces and Resources Policy Center, contact the Director, James Hosek. He can be reached by email at james_hosek@rand.org; by phone at 310-393-0411, extension 7183; or by mail at the RAND Corporation, 1776 Main Street, Santa Monica, CA 90407-2138. More information about RAND is available at www.rand.org.

Contents

Figures

Tables

Summary

Several recent studies have suggested the need for the Department of Defense (DoD) to revisit joint manpower matters and develop a strategic approach to Joint Officer Management (JOM) and Joint Professional Military Education (JPME). For example, an independent study[1] authorized under the 2002 National Defense Authorization Act (NDAA) indicated that JOM and JPME require updating in practice, policy, and law to meet the demands of a new era more effectively and that such changes should be undertaken as part of an overall strategic approach to developing the officer corps for joint warfare.

In fiscal year (FY) 2003, DoD asked the RAND National Defense Research Institute (NDRI) to undertake an analysis that was intentionally broad, one that would look beyond manpower issues to establish the context for officer development in joint matters. The results of that earlier effort are documented in Thie et al. (2005), which presents the findings of the joint officer analysis, a conceptual strategic approach for joint officer management, and recommendations for operationalizing the strategic plan.

This current work builds on that earlier effort. While the overall project has broader goals, its principal goal is to operationalize the strategic approach for joint officer management in the active component through extensive data analysis and complex modeling. As a lead-in to this effort, the research sponsor and another organization conducted a Web-based survey of individuals serving in billets that were likely to either require prior joint experience or joint education or provide officers with joint experience. The Joint Officer Management (JOM) Census survey (or JOM survey) was conducted in summer 2005. Surveyed billets include those currently on the Joint Duty Assignment List (JDAL), non-JDAL billets in external organizations that have some billets on the JDAL, and non-JDAL internal service billets nominated by the services as providing joint experience or requiring joint experience or joint education.

This report provides an overview of the survey responses. It examines officers' answers regarding their assignments, including the extent to which they believe their assignments provide them with joint experience or require them to have had prior joint education, training, or experience and whether and how these answers differ across organizations and/or services where the billets are located.

This study is designed to set the stage for future, more complex analysis. The next phase of this project will examine the extent to which prior "jointness" is required by billets and whether sufficient numbers of officers with joint education, training, and experience are likely to be available to satisfy that requirement.

---

[1] Booz Allen Hamilton, 2003.

Survey Methodology

The JOM survey addresses the following types of billets:

- Billets currently on the JDAL[2]
- Non-JDAL billets in external organizations that have some billets on the JDAL
- Internal service billets not on the JDAL that were nominated by the services as providing joint experience or requiring joint experience or joint education.

Services and external organizations were asked to identify (1) billets for which a prerequisite joint education course and/or experience gained through a previous joint tour of duty might better qualify an officer to perform the mission requirements of his or her position and (2) billets that provide officers with significant experience in joint matters (for example, billets that provide incumbents with multinational, multiservice, or interagency experience) and thus could be deemed to be similar to joint duty assignments. Each service used its own criteria for nominating billets.

Although the intention had been to survey all billets that met the criteria above—in other words, to conduct a census of actual and potential JDA billets—the survey actually encompassed a subset of billets rather than the census because of an outdated sampling frame and some inadvertently excluded organizations. The 30,043 billets that were surveyed included 8,475 JDAL billets (out of 9,700 billets in 2004); 6,384 billets in external organizations (which encompassed almost all the billets in most of the major billet organizations with some exceptions, most notably intelligence organizations); and 15,184 service-nominated billets (which presumably covered all the billets the services designated as meeting their criteria). We received a total of 21,214 responses—a response rate of 71 percent. However, the response rates varied considerably across billet organizations.

Categorization of Billets

We used two major classification schemes to examine the differences in the responses:

- **JDAL Status:** Billets are categorized into one of three groups: billets currently on the JDAL; non-JDAL billets in external organizations with some billets on the JDAL; internal service billets (which are not on the JDAL by law) that were nominated by the four services.
- **Major Billet Organization**: Billets are categorized according to the organization in which the billet is currently assigned. Those mutually exclusive groups include the following:

[2] A Joint Duty Assignment (JDA) is an assignment to a billet in a multiservice or multinational command/activity that is involved in the integrated employment or support of the land, sea, and air forces of at least two of the three Military Departments. The Joint Duty Assignment List is a consolidated roster that contains all billets that are approved JDAs for which joint credit can be applied. Billets are added to and deleted from the JDAL, and there is a validation process to review positions nominated for addition. Because of problems with the sampling frame, not all JDAL billets were included in the survey.

- U.S. Army
- U.S. Navy
- U.S. Air Force
- U.S. Marine Corps
- Joint Staff
- Office of the Secretary of Defense (OSD)
- U.S. Central Command joint task force (CENTCOM JTF)
- International organizations
- Combat support agencies (CSAs)
- Other non-OSD defense agencies
- OSD defense agencies
- Educational agencies
- Geographic commands
- Force provider
- Functional commands.[3]

About half the billets in the responding sample had been nominated by the services, 29 percent were JDAL billets, and the remaining billets were non-JDAL billets in external organizations. Together, the Army and Air Force billets account for 40 percent of the responding sample. Navy billets account for a little less than 10 percent of the sample, and the Marine Corps accounts for 0.3 percent of the sample.

Caveats

Sample sizes are quite small for some groups; thus, the findings should be seen as suggestive rather than definitive. Because we are unable to correct for nonresponse, it is important to view the findings here as being representative of the responding sample and not the entire universe of joint or potentially joint officers.

Although we show differences in characteristics of the billets nominated by the four services, it is important to remember that these billets cannot and should not be directly compared. The services were provided with broad criteria for nominating billets; however, how the criteria should be operationalized and any additional criteria to be used were left up to the individual services. As a result, the billets nominated by the services are wide-ranging, with the Marine Corps being the most selective and most parsimonious in its nominations. Thus, one should not expect these billets to be comparable or draw inferences regarding how "joint" billets in one service are compared with those in another.

Typical Metrics of "Jointness"

The JOM survey gathered information by asking questions about a number of billet characteristics that are generally regarded as defining jointness—types of tasks performed during a

---

[3] See Table 3.1 in Chapter Three for a crosswalk between organizations that were surveyed and the categorization used here.

typical workweek; supervision of the billet by non–own-service or civilian personnel; frequency and number of interactions with non–own-service organizations and personnel; the need for joint professional education or prior joint experience for successful job performance; and types of joint experience provided by the billet. There are other measures of jointness, but these characteristics are a reasonable subset to use for purposes of this analysis. We use officers' responses to these questions to provide a broad-brush picture of how billets in various organizations rank along the various dimensions.

Table S.1 defines the set of indicators used to characterize jointness. Tables S.2 through S.4 describe the billets along the various dimensions by JDAL category and billet organization.

Tasks Performed During a Typical Workweek

Four tasks were selected as representing "highly joint" activities—(1) providing strategic direction and integration; (2) developing/assessing joint policies; (3) developing/assessing joint doctrine; and (4) fostering multinational, interagency, or regional relations. Officers were much more likely to report doing the first task than the other three, as shown in Table S.2. Close to 80 percent of JDAL officers performed one or more of these tasks,[4] and 27 percent

Table S.1
Definitions of Indicators Used to Characterize "Jointness"

Metric	Indicator
Tasks performed during the typical workweek	Percentage of officers providing strategic direction and integration
	Percentage of officers developing/assessing joint policies
	Percentage of officers developing/assessing joint doctrine
	Percentage of officers fostering multinational, interagency, or regional relations
	Percentage of officers performing three or more of these tasks
Interactions with non–own-service organizations and personnel	Median number of non–own-service organizations with whom officers interact monthly or more frequently
	Median number of non–own-service personnel with whom officers interact monthly or more frequently
Supervision of billet by non–own-service personnel/civilians	Percentage of officers reporting being supervised by one or more non–own-service supervisor/civilian/non–U.S. military personnel or civilian
Need for joint professional education or prior joint experience	Percentage of officers reporting that JPME II is required or desired for the assignment
	Percentage of officers reporting that prior joint experience is required or desired for the assignment
Types of joint experience provided by the billet	Percentage of officers reporting getting significant experience in multiservice matters
	Percentage of officers reporting getting significant experience in multinational matters
	Percentage of officers reporting getting significant experience in interagency matters
	Percentage of officers reporting getting significant experience in all three areas

[4] See Chapter Five for further details.

Table S.2
Rankings of Billet Categories Based on Tasks Performed During the Typical Workweek

	Percentage of Officers Providing Strategic Direction and Integration	Rank	Percentage of Officers Developing or Assessing Joint Policies	Rank	Percentage of Officers Developing or Assessing Joint Doctrine	Rank	Percentage of Officers Fostering Multinational, Interagency, or Regional Relations	Rank	Percentage of Officers Performing Three or More of These Tasks	Rank
JDAL Category										
JDAL billets	59.0	1	37.5	1	32.8	1	31.6	1	27.0	1
Non-JDAL billets in external organizations	45.2	2	22.8	2	18.8	2	21.3	2	14.6	2
Service-nominated billets	34.0	3	12.1	3	12.9	3	15.0	3	8.5	3
Major Billet Organization										
Joint Staff	68.3	2	54.4	1	44.1	2	21.6	8	36.4	1
OSD	77.6	1	45.2	2	22.8	8	24.7	6	25.5	5
CENTCOM JTF	42.4	12	18.3	12	11.2	15	33.5	3	10.4	12
Army	26.4	15	10.2	15	12.7	13	17.1	11	8	14
Navy	33.1	14	11.8	14	11.5	14	16.4	12	7.9	15
Air Force	40.7	13	13.6	13	13.1	12	12.4	13	9	13
Marine Corps	55.4	5	36.9	6	41.5	4	20.0	9	23.1	8
International organizations	47.1	10	33.6	7	25.4	7	48.8	1	24.6	6
CSAs	46.3	11	18.8	11	14.8	11	25.1	5	12	11
Other non-OSD defense agencies	54.7	6	25.4	10	21.4	9	27.5	4	20.5	9
OSD defense agencies	57.1	4	31.7	9	20.5	10	9.3	14	18	10
Educational agencies	49.8	9	40.5	4	43.5	3	24.1	7	30.4	2
Geographic commands	52.5	7	32.1	8	28.6	6	34.5	2	23.6	7
Force provider	50.7	8	40.1	5	46.0	1	19.6	10	28.2	4
Functional commands	62.3	3	41.0	3	38.6	5	8.7	15	28.3	3

NOTES: This table summarizes data found in Tables 5.1 and 5.4 in Chapter Five.
Organizations with equal percentages of officers meeting the particular criterion were given the same rank.

of JDAL officers performed at least three of these tasks. In contrast, only 45 percent of officers in internal service billets performed any of these joint tasks, and less than 10 percent of officers in internal service billets performed three or more of these tasks. The non-JDAL, non-service billets ranked in the middle, with 60 percent of officers performing at least one of the four tasks and 15 percent performing at least three of the tasks. Eighty-five percent of officers in Joint Staff or OSD staff billets and 75 percent of officers in Marine Corps, international organization, force provider, and functional command billets performed one or more of these tasks. Between 22 and 36 percent of officers at these organizations performed three or more of these tasks.

Based on this set of indicators, JDAL and non-JDAL billets in external organizations rank either first or second among JDAL categories, while internal service billets rank third. When we examine major billet organizations, billets assigned to the Joint Staff, OSD staff, educational agencies, force provider, and the functional commands seem to rank high on these measures of "jointness." The rankings of the others are more mixed, with some ranking high on tasks performed and others ranking high on the "providing strategic direction and integration" indicator. The Marine Corps billets rank higher than the other three services on all four measures of jointness.

Frequency and Number of Interactions with Non–Own-Service Organizations and Personnel

Officers in JDAL billets tended to interact frequently with the highest number of non–own-service organizations (six) while officers assigned to internal service billets interacted with one (see Table S.3). Officers serving in the OSD or Joint Staff reported interacting with between nine and 13 non–own-service organizations frequently. The defense agencies, combat support agencies, and the combatant commands reported interacting with five non–own-service organizations. CENTCOM JTF, educational agency, and international organization billets interacted frequently with two to three organizations. Among the services, officers in Marine Corps billets interacted with three non–own-service organizations compared with one for the Army and Air Force.

Overall, using the median, JDAL and non-JDAL billet officers reported interactions with five types of personnel, compared with two for officers serving in the billets nominated by the services. Officers serving in CENTCOM JTF, educational agencies, and geographic commands reported interacting frequently with six types of personnel (excluding own-service personnel), while all other non-service billets reported interacting with five types of non–own-service personnel. Among the services, the median for the Navy and Marine Corps billets was four types of personnel compared with two for the Army and Air Force billets.

Supervision of Billet by Non–Own-Service Personnel

Not unexpectedly, close to 80 percent of JDAL billets and about 75 percent of non-JDAL billets in external organizations are supervised by at least one non–own-service supervisor, compared with a little more than 20 percent of the service-nominated billets (see Table S.3).

Officers serving in OSD staff or OSD defense agency billets are almost all supervised by at least one non–own-service supervisor, and this was true of 75 percent or more of officers assigned to other agencies, the Joint Staff, the combatant commands. Only 45 percent of

Table S.3
Rankings of Billet Categories Based on Frequent Interactions with Non–Own-Service Organizations and Personnel and Non–Own-Service Supervision

	Median Number of Non–Own-Service Organizations with Whom Officers Interact Monthly or More Frequently	Rank	Median Number of Non–Own-Service Personnel with Whom Officers Interact Monthly or More Frequently	Rank	Percentage of Billets Supervised by Non–Own-Service Supervisors or Civilians	Rank
JDAL Category						
JDAL billets	6	1	5	1	78	1
Non-JDAL billets in external organizations	4	2	5	1	75.4	2
Service-nominated billets	1	3	2	3	21	3
Major Billet Organization						
Joint Staff	13	1	5	4	77.8	8
OSD	9	2	5	4	94.6	2
CENTCOM JTF	3	9	6	1	50.1	10
Army	1	14	2	14	8.9	15
Navy	2	12	4	12	29.6	12
Air Force	1	14	2	14	26.8	13
Marine Corps	3	9	4	12	13.8	14
International organizations	2	12	5	4	45.5	11
CSAs	5	3	5	4	83.9	4
Other non-OSD defense agencies	5	3	5	4	83.7	6
OSD defense agencies	5	3	5	4	96.1	1
Educational agencies	3	9	6	1	85.5	3
Geographic commands	5	3	6	1	74.5	9
Force provider	5	3	5	4	83.9	5
Functional commands	5	3	5	4	79	7

NOTES: This table summarizes data found in Tables 4.4, 4.6, and 4.7 in Chapter Four and Figures 3.2 and 3.3 in Chapter Three. Organizations with the same median number or equal percentages of officers meeting the particular criterion were given the same rank.

those serving in international organizations reported having at least one non–own-service supervisor. Of the major billet organizations, the four services, particularly the Army and Marine Corps, were the least likely to have supervisors from other organizations.

JDAL billets rank first on both indicators (interaction with non–own-service organizations and interaction with non–own-service personnel); Joint Staff and OSD billets rank first or second. However, CENTCOM JTF billets rank low on the interactions-with-organizations indicator but rank very high on the interactions-with-personnel indicator. Other non-service organizations were in the middle, with the services generally ranking last on these indicators, with the exception of the Marine Corps.

Need for Joint Professional Education and Prior Joint Experience for Billet Assignment

The majority of officers believed that joint professional education (JPME II) and prior experience in a joint environment were required or desired to perform their duties successfully (see Table S.4).[5] Officers in internal service billets were less likely to report as such, but even among those officers, between 70 and 80 percent believed that such education and experience is required or desired for effective job performance. Among major billet organizations, well over 80 percent of officers in non-service billets and in the Marine Corps billets reported that joint education and experience were required or desired for the assignment.

Although the ranking reflects these percentages, even among those organizations that ranked last, 60–78 percent of officers reported a need for joint education and experience.

Joint Experience Provided by a Billet

JDAL billets provided the most experience in multiservice, multinational, and interagency matters (see Table S.5). Overall, 87 percent of officers in JDAL billets reported that they gained significant experience in multiservice matters, and between 65 and 75 percent reported gaining significant experience in multinational and interagency matters. Officers in

Table S.4
Rankings of Billet Categories Based on Need for Joint Professional Education and Prior Joint Experience in Billet Assignment

	Percentage of Officers Reporting that JPME II Is Required or Desired for the Assignment	Rank	Percentage of Officers Reporting that Prior Joint Experience Is Required or Desired for the Assignment	Rank
JDAL Category				
JDAL billets	91.7	1	88.9	1
Non-JDAL billets in external organizations	86.3	2	84.9	2
Service-nominated billets	70.9	3	69.9	3
Major Billet Organization				
Joint Staff	90.4	5	86.7	5
OSD	91.8	4	90.5	3
CENTCOM JTF	84.7	10	85.3	7
Army	77.7	13	73.7	13
Navy	60.7	15	64.4	15
Air Force	69.6	14	69.1	14
Marine Corps	82.5	12	83.6	9
International organizations	85.1	9	83.1	11
CSAs	84.3	11	84.9	8
Other non-OSD defense agencies	88.1	8	83.2	10
OSD defense agencies	89.9	6	86.3	6
Educational agencies	95.2	1	95.0	1
Geographic commands	93.3	3	90.0	4
Force provider	94.2	2	91.0	2
Functional commands	88.4	7	82.8	12

NOTES: This table summarizes data found in Figures 6.2 and 6.3 and Table 6.10 in Chapter Six. Organizations with equal percentages of officers meeting the particular criterion were given the same rank.

[5] Large percentages of officers reported that they had no experience with JPME II. For example, 52 percent of officers in non-JDAL billets in external organizations indicated that they had no experience with JPME II, as did 59 percent of officers in internal service billets. We restricted the responses to those with experience with JPME II when calculating the percentages reporting that JPME II was required or desired.

Table S.5
Rankings of Billet Categories Based on Types of Joint Experience Provided by the Billet

	Percentage of Officers Reporting Getting Significant Experience in Multiservice Matters	Rank	Percentage of Officers Reporting Getting Significant Experience in Multinational Matters	Rank	Percentage of Officers Reporting Getting Significant Experience in Interagency Matters	Rank	Percentage of Officers Reporting Getting Significant Experience in All Three Areas	Rank
JDAL Category								
JDAL billets	86.9	1	65.0	1	75.1	2	53.4	1
Non-JDAL billets in external organizations	79.1	2	56.7	2	75.9	1	47.5	2
Service-nominated billets	48.9	3	39.6	3	43.8	3	23.5	3
Major Billet Organization								
Joint Staff	89.3	2	57.1	5	74.4	7	51.6	4
OSD	92.5	1	55.8	6	84.6	2	50.6	5
CENTCOM JTF	82.4	7	81.2	2	71.9	8	63.1	2
Army	49.7	14	50.4	10	42.1	15	28.5	11
Navy	52.3	13	37.2	12	43.6	14	23.8	14
Air Force	46.2	15	31.3	14	45.7	12	19.1	15
Marine Corps	77.1	10	45.9	11	44.3	13	24.6	13
International organizations	65.9	12	96.3	1	47.8	11	39.5	9
CSAs	77.8	9	54.3	8	81.3	5	45.8	7
Other non-OSD defense agencies	75.8	11	55.4	7	81.6	4	48.3	6
OSD defense agencies	86.4	4	36.6	13	85.1	1	34.8	10
Educational agencies	81.4	8	71.7	4	83.1	3	65.0	1
Geographic commands	88.7	3	73.2	3	76.7	6	59.8	3
Force provider	85.9	5	51.5	9	62.1	10	45.8	8
Functional commands	82.7	6	30.5	15	62.7	9	26.5	12

NOTES: This table summarizes data found in Figures 6.7 through 6.12 and Table 6.10 in Chapter Six. Organizations with equal percentages of officers meeting the particular criterion were given the same rank.

non-JDAL, non-service billets were much less likely to report gaining experience with multinational matters than with multiservice or interagency matters. Compared with officers in non-service billets, officers in service-nominated billets were less likely to report gaining experience in these areas

More than 70 percent of the non-service organizations provide significant experience in multiservice matters, and this is also true of Marine Corps billets. By comparison, 46–52 percent of the Army, Navy, and Air Force billets provide such experience. Other billets that ranked high on this indicator were educational agency, OSD defense agency, geographic command, and force provider billets.

Almost all officers assigned to international organizations reported getting significant experience in multinational matters. Other organizations that ranked high on this indicator include CENTCOM JTF, geographic commands, and educational agencies, with 70–80 percent of officers in these organizations agreeing or strongly agreeing that their billets provide significant experience in multinational matters.

With a few exceptions, non-service billets provide significant amounts of experience in interagency matters. Those serving in OSD staff or other agency billets were particularly likely to agree strongly with this statement. About 40–45 percent of officers in internal service billets reported getting such experience.

Ninety percent or more of JDAL and non-JDAL, non-service billets provide experience in at least one of the three joint areas (multiservice, multinational, or interagency). About half provide experience in all three areas, compared with 24 percent of internal service billets. Well over 85 percent of billets, except for those in the services, provide significant experience in at least one of these areas, and, with some exceptions, well over 70 percent provide significant experience in two of the areas. More than half of the billets in the educational agencies, CENTCOM JTF, geographic commands, Joint Staff, and OSD staff provide significant experience in all three areas.

Not unexpectedly, JDAL billets rank first on every indicator. Educational agencies, CENTCOM JTF billets, and billets in the geographic commands rank very high on providing significant experience in all three areas.

Conclusions

The 2005 Joint Officer Management Census survey was designed to elicit information on joint billets on the JDAL, potential joint billets in external organizations with some billets on the JDAL, and internal service billets nominated by the services as requiring or providing joint experience. The findings provide a rich, descriptive portrait of the experiences of officers in the various joint or potential joint billets. In Chapter Eight, we show how the data can be used to support more-analytical models of joint officer supply and demand: (1) to estimate numbers of billets available to provide joint duty assignments based on experiences afforded by the billets and (2) to estimate demand for joint duty billets based on reported need for joint education, joint education, or both.

However, grouping the billets as we have done for ease of exposition and to portray overall trends hides important information. In deciding whether a particular billet provides joint experience or requires joint experience, training, and/or education—and as such deserves recognition as a joint duty assignment—the billet ultimately needs to be examined on its own merits. That is the purpose of the next phase of this project. Continuing work by RAND will examine the extent to which prior jointness is required by billets and whether sufficient numbers of officers with joint education, training, and experience are likely to be available to satisfy that requirement.

Acknowledgments

This research could not have been carried out without the participation of the several thousand respondents who generously gave of their time to respond to the survey that is the subject of this report. We appreciate their efforts and their dedication to improving the joint officer management system by providing detailed information on their experiences and perceptions. We are grateful for the support of Lieutenant Colonel Charles Armentrout from our sponsoring office, and we thank the staff from LMI who worked patiently to answer our many questions regarding the survey.

At RAND, we thank our reviewers, Scott Naftel and Peter Schirmer, for their constructive and very helpful reviews and for completing the reviews under tight time constraints. Their suggestions contributed greatly to the substance and clarity of the report. Heather Barney, Sarah Montague, and Luetta Pope provided invaluable help in checking the data and in formatting an earlier version of this report. We also thank our editor, Nancy DelFavero, for her impeccable editing and for her patience, good humor, and thoroughness.

Acronyms and Abbreviations

AAV	amphibious assault vehicle
AFE	Air Forces, Europe
AFR	Air Force Reserve
AFSC	Armed Forces Specialty Code
AMC	Army Materiel Command
AOC	area of concentration
AOR	area of responsibility
C4	command, control, communications, computers,
C4I	command, control, communications, computers and intelligence
C4ISR	command, control, communications, computers, intelligence, surveillance, and reconnaissance
CBRNE	chemical, biological, radiological, nuclear, explosive
CENTCOM	U.S. Central Command
CJCS	Chairman of the Joint Chiefs of Staff
CJTF	Combined Joint Task Force
CJTF-AFG	CJTF—Afghanistan
CJTF-HOA	CJTF—Horn of Africa
CJTF-7	CJTF—Iraq
COCOM	Combatant Command
COS	Critical Occupation Specialty (officer)
CSA	combat support agency
DHS	Department of Homeland Security
DIA	Defense Intelligence Agency
DISA	Defense Information Systems Agency
DoD	Department of Defense
DTRA	Defense Threat Reduction Agency
ELC	Electronic Security Command
ELM	Air Force Elements
EOCOM	U.S. European Command
FEMA	Federal Emergency Management Agency

FSA	Family Separation Allowance
FY	fiscal year
GAO	Government Accountability Office (formerly the General Accounting Office)
GNA	Goldwater-Nichols Act
HAF	Headquarters Air Force
HQ	headquarters
IADB	Inter-American Defense Board
ICAF	Industrial College of the Armed Forces
IO	information operations
IT	information technology
IW	information warfare
JAWS	Joint Advanced Warfighting School
JCS	Joint Chiefs of Staff
JCWS	Joint and Combined Warfighting School
JDA	Joint Duty Assignment
JDAL	Joint Duty Assignment List
JFC	joint forces commander
JFCOM	U.S. Joint Forces Command
JFSC	Joint Forces Staff College
JOM	Joint Officer Management
JPME	Joint Professional Military Education
JSO	joint specialty officer
JTF	joint task force
MOOTW	military operations other than war
MOS	Military Occupational Specialty
NATO	North Atlantic Treaty Organization
NCA	National Command Authority
NDAA	National Defense Authorization Act
NDRI	RAND National Defense Research Institute
NDU	National Defense University
NGO	nongovernmental organization
Nom	nominated (officer)
NORTHCOM	U.S. Northern Command
NSA	National Security Agency
NSC	National Security Council
NWC	National War College
OSD	Office of the Secretary of Defense
PACOM	U.S. Pacific Command
PAF	Pacific Air Forces

PAJE	Process for Accreditation of Joint Education
PDUSD (P&R)	Principal Deputy Under Secretary of Defense, Personnel, and Readiness
R&D	research and development
ROK	Republic of Korea
SecDef	Secretary of Defense
SOCOM	U.S. Special Operations Command
SOUTHCOM	U.S. Southern Command
SPACECOM	Space Command
SPAWAR	Space Warfare Command
STRATCOM	U.S. Strategic Command
TDY/TAD	temporary duty travel/temporary additional duty
TRADOC	Training and Doctrine Command
TRANSCOM	U.S. Transportation Command

CHAPTER ONE

Introduction

Background and Purpose of Study

The Goldwater-Nichols Act (GNA) of 1986[1] forged a cultural revolution in the U.S. armed forces by improving the way in which the Department of Defense (DoD) prepares for and executes its mission. Title IV of the GNA addresses joint officer personnel policies and provides specific personnel management requirements for the identification, education, training, promotion, and assignment of officers to joint duties. Provisions in the act directed the Secretary of Defense (SecDef) to develop a definition of Joint Duty Assignment (JDA) and to publish a Joint Duty Assignment List (JDAL).[2] This list includes those positions at organizations outside the individual services that address issues involving multiple services or other nations where an assigned officer gains "significant experience in joint matters."[3] Joint duty consideration was limited to pay grades O-4 or higher. All such positions in some organizations (Office of the Secretary of Defense, the Joint Staff, and the unified commands) and half of the positions in each defense agency were placed on the JDAL. Assignments within an officer's own military department and assignments for joint education or training were specifically excluded from the definition of joint duty assignments.

Several studies[4] have suggested the need for DoD to revisit joint manpower matters and develop a strategic approach to Joint Officer Management (JOM) and Joint Professional Military Education (JPME).[5] For example, a study by the General Accounting Office (GAO) (renamed the Government Accountability Office in July 2004) conducted an assessment of DoD's actions to implement provisions in the law that address the development of officers in joint matters. The study concluded that "a significant impediment affecting DoD's ability to fully realize the cultural change that was envisioned by the act is the fact that DoD has not taken a strategic approach to develop officers in joint matters."[6] In March 2003, an inde-

[1] *Goldwater-Nichols Department of Defense Reorganization Act of 1986,* Public Law 99-433, October 1, 1986.

[2] A Joint Duty Assignment is an assignment to a billet in a multiservice or multinational command/activity that is involved in the integrated employment or support of the land, sea, and air forces of at least two of the three Military Departments. The Joint Duty Assignment List is a consolidated roster that contains all billets that are approved JDAs for which joint credit can be applied. Billets are added to and deleted from the JDAL, and there is a validation process to review positions nominated for addition. Because of problems with the sampling frame, not all JDAL billets were included in the survey.

[3] Title 10, Section 668 of the U.S. Code (USC) specifically excludes "(A) assignments for joint training or joint education; and (B) assignments within an officer's own military department" from the definition of joint duty assignments.

[4] General Accounting Office, 2002; Booz Allen Hamilton, 2003.

[5] Joint Professional Military Education is defined later in this chapter.

[6] General Accounting Office, 2002.

pendent study[7] authorized under the 2002 National Defense Authorization Act (NDAA) indicated that JOM/JPME requires updating in practice, policy, and law to meet the demands of a new era more effectively and that such changes should be undertaken as part of an overall strategic approach to developing the officer corps for joint warfare.

In fiscal year (FY) 2003, DoD asked the RAND National Defense Research Institute (NDRI) to undertake an analysis that was intentionally broad, one that would look beyond manpower issues to establish the context for officer development in joint matters. The intent of such a strategic approach is to provide overarching guidance on officer training and development in joint matters to best meet DoD's mission and goals in the context of evolving combatant commander and service requirements, revolutionary changes in technology, and a dramatic cultural shift in the military that require the services to move from differentiation to integration of their workforces. Thus, a strategic approach to human resource management determines which critical workforce characteristic or characteristics are needed given missions, goals, and desired organizational outcomes; assesses the availability of the characteristic(s) now and in the future; and suggests changes in management practices for personnel with those characteristic(s) to minimize gaps between need and availability. The results of that earlier effort are documented in Thie et al. (2005), which presented the findings of the joint officer analysis, a conceptual strategic approach for joint officer management, and recommendations for operationalizing the strategic plan. That report also pointed out that the next research step to operationalize or implement the strategic plan for joint officer management was to gather extensive data on billets that require joint experience, education, or training and billets that provide such experience, and it outlined a detailed plan for gathering the data through a Web-based survey.

This current work builds on that earlier effort. While the overall project has broader goals, the principal goal of the project is to operationalize the strategic approach for joint officer management in the active component through extensive data analysis and complex modeling. As a lead-in to this effort, the research sponsor and another organization gathered the data thought to be necessary to operationalize the framework through a Web-based survey of individuals serving in billets that were likely to either require prior joint experience or provide officers with joint experience. The Joint Officer Management (JOM) Census survey (hereafter the JOM survey) was conducted in summer 2005. Surveyed billets include those currently on the Joint Duty Assignment List, non-JDAL billets in external organizations that have some billets on the JDAL, and non-JDAL internal service billets nominated by the services. The next phase of the project is examining the extent to which prior "jointness" is required by billets and whether sufficient numbers of officers with joint education, training, and experience are likely to be available to satisfy that need.

Purpose of This Report

This report provides an overview of the approximately 21,000 responses to the Joint Officer Management Census survey conducted in summer 2005. It is designed to set the stage for the more complex task of analytic job evaluation. It examines officers' answers on their backgrounds and assignments, including the extent to which officers believe that their assign-

[7] Booz Allen Hamilton, 2003.

ments provide them with joint experience or require them to have had prior joint education, training, or experience and whether and how these answers differ across the organizations and/or services in which these billets are located. The primary purpose of this report is to provide readers with a comprehensive reference source for the data collected in the JOM survey and to foreshadow how the data could be used to anchor a strategic approach to joint officer management. The report also offers insights into some of the critical personnel issues regarding joint officer management raised in Thie et al. (2005). These issues include whether other billets not currently on the JDAL provide a joint experience rich enough to be considered eligible for granting joint credit; the relative value of joint professional education and training currently required (with some exceptions) to become a joint specialty officer; and the award of joint duty credit for serving on joint task forces (JTFs), either in JTF headquarters or in service component commands and service units assigned to JTFs.

Caveats

Because the response rates varied markedly across organizations, the sample sizes are quite small for some groups; thus, the findings should be seen as suggestive rather than definitive. In addition, because we are unable to correct for nonresponse, and are somewhat unclear as to the level of nonresponse for some organizations, the findings should be viewed as being representative of the responding sample and not the entire universe of joint or potentially joint officers. Despite these caveats, the survey offers a rich and complex look at the world of joint officers, the nature of their assignments, their typical workloads, their interactions, and their perceptions of the value of training, education, and experience.

Although we show differences in the characteristics of billets nominated by the four services, it is important to remember that these billets cannot and should not be directly compared. The services were provided with broad criteria for nominating billets; however, how the criteria should be operationalized and any additional criteria to be used were left up to the individual services. As a result, the billets nominated by the services are wide-ranging, with the Marine Corps being the most selective and most parsimonious in its nominations. Thus, one should not expect these billets to be comparable or draw inferences regarding how "joint" billets in one service are compared with those in another.

Terminology[8]

Joint Duty Assignment

A JDA is an assignment to a billet in a multiservice or multinational command/activity that is involved in the integrated employment or support of the land, sea, and air forces of at least two of the three Military Departments.[9] The duties of an officer in qualifying for a JDA involve producing or promulgating national military strategy, joint doctrine, joint policy, strategic plans or contingency plans or commanding and controlling operations under a combat-

[8] This section is adapted from Thie et al., 2005.

[9] DoD Instruction (DoDI) 1300.20, 1996.

ant command. Assignments to an officer's own Military Department or assignments for joint education or training do not qualify and are not covered by this definition. Successful completion of a JDA is a criterion for designation as a JSO.

Joint Duty Assignment List

As stated earlier, the JDAL is a consolidated roster that contains all billets[10] that are approved JDAs for which joint credit can be applied. Billets are added to and deleted from the JDAL, and there is a validation process to review positions nominated for addition. A joint duty validation board composed of representatives of the Office of the Assistant Secretary of Defense, the Joint Staff, and the Military Departments consider the joint content of nominated billets. A billet is evaluated and voted on by its merit for inclusion in or exclusion from the JDAL by the Validation Board.

Critical Joint Duty Assignments

These assignments are JDA positions in which either the incumbent should be experienced and educated in joint matters, or the position would be greatly enhanced by an officer with the joint experience and education. Critical positions are proposed by heads of joint activities, approved by the Principal Deputy Under Secretary of Defense, Personnel and Readiness (PDUSD (P&R)) with the advice and assistance of the Chairman of the Joint Chiefs of Staff (CJCS) and documented in the JDAL. Critical positions are to be filled by JSOs unless waived by the CJCS.[11] Like the JDAL, critical billets also change over time.

Critical Occupation Specialty (COS) Officer

A COS is a military occupational field that involves combat operations within the services and in which the Secretary of Defense has determined that a shortage of trained officers exists. Specialties that may be designated "COS" are determined each year. For example, in 2004, the designated COSs for the services (per the Annual Defense Report for 2004)[12] were:

- Army—infantry, armor, artillery, air defense artillery, aviation, special operations, combat engineers
- Navy—surface, submariner, aviation, SEALS, special operations
- Air Force—pilot, navigator, command/control, operations, space/missile operations
- Marine Corps—infantry, tanks/amphibious assault vehicles (AAVs), artillery, air control/air support, anti-air warfare, aviation, and engineers.

The Secretary of Defense may reduce the JDA tour lengths of COS officers to two years, as long as such "COS takeouts" do not exceed 12.5 percent of the officers serving as JSOs and JSO nominees.

---

[10] In the past, only 50 percent of the positions in defense agencies could qualify as JDAs, whereas 100 percent of the positions in other joint organizations were on the JDAL. These limits no longer exist.

[11] DoDI 1300.20, 1996.

[12] Rumsfeld, 2004.

Joint Duty Credit

Joint credit is granted for the completion of an assignment (or accumulation of sufficient time in assignments) to a JDA that meets all statutory requirements.[13] As mentioned above, there are two types of positions on the JDAL: the standard joint duty position and joint critical positions. Any qualified officer may serve in the standard JDA,[14] while only fully qualified JSOs may fill the critical joint duty assignments.

Joint Professional Military Education

JPME focuses specifically on joint matters. The National War College (NWC) and Industrial College of the Armed Forces (ICAF) course of instruction provides full JPME Phase I and JPME Phase II (hereafter in this report JPME I and JPME II) credit for graduates. In addition, JPME I instruction is provided at the Military Service Colleges (resident and non-resident) at the intermediate level (O-4) and senior level (O-5/O-6) and at the National Defense University (NDU).

In addition to JPME credit from NWC and ICAF, the Joint Forces Staff College (JFSC) provides JPME II[15] to both intermediate and senior-level students. A small number of candidates take JPME II at the Joint and Combined Warfighting School (JCWS).

Intermediate Level Colleges teach joint operations from the standpoint of service forces in a joint force supported by service component commands. Senior-level Military Service Colleges address theater-level and national-level strategies and processes. Curricula focus on how unified combatant commanders, the Joint Staff, and DoD use the instruments of national power to develop and carry out national military strategy.[16]

The JCWS at the JFSC (for JPME II credit) provides instruction in joint operations from the perspective of the CJCS, Joint Chiefs of Staff, unified combatant commanders, and JTF commanders. The course develops joint attitudes and perspectives and exposes officers to other service cultures while maintaining a concentration on Joint Staff operations.

To meet the educational prerequisites to become a JSO or JSO nominee, officers must at a minimum complete one of the following:

- JPME I at (an accredited) intermediate- or senior-level Military Service College.[17]
- JPME II at NWC, ICAF, JFSC, and JCWS.

As of next year, senior-level service programs will become eligible for future Process for Accreditation of Joint Education (PAJE) accreditation for JPME II. These include the U.S. Army War College, U.S. Navy College of Naval Warfare, U.S. Marine Corps War College, and U.S. Air Force Air War College.

---

[13] Chairman of the Joint Chiefs of Staff Instruction 1331.01B, 2003.

[14] Additionally, there are provisions for awarding of joint creditable service for duty performed in approved Joint Task Force Headquarters assignments. Constraints and limitations are applicable to each period of creditable service, and waivers can be applied, with certain provisions.

[15] The JPME II course of instruction at JFSC was recently reduced in duration from 12 weeks to 10 weeks, which allows for an additional session to be held each year. Liaison with JFSC officials indicates that four sessions are now held (beginning in FY 2005), with the maximum capability of 255 students per session.

[16] CJCS Instruction 1800.01B, 2004.

[17] Officers (other than those with a critical occupational specialty) must attend JPME II prior to completing their joint assignment to qualify as a JSO. Attendance at JPME II prior to completing JPME I requires a waiver by CJCS.

JSO Designation

While there is no longer a board process for JSO selection, officers who complete JPME II and JDAs are nominated to the Office of the Secretary of Defense (OSD) for JSO designation. There are four categories of officers who are considered for JSOs:

- Officers who completed JPME I and JPME II and a full JDA (36 months), or COS officers who completed a 36-month JDA before completing JPME I and JPME II
- COS officers who completed JPME I and JPME II and a 22–24 month tour[18]
- Other officers who complete a JDA prior to JPME II (requires a waiver)
- Officers who complete two full JDAs but no JPME II.

The total number of waivers granted for officers for a fiscal year may not exceed 10 percent of the total number of officers in that pay grade selected for the joint specialty during that fiscal year.

Joint Task Force

JTFs are the primary organizations for joint operations and capitalize on the capabilities of each service. Many JTFs are designed to accomplish specific objectives and then disestablish. Others are permanent or long-standing. Service component commands are frequently used as the basis for them; other options are standing JTF headquarters or formation of an ad hoc headquarters from various contributors. The fluidity of these organizations was not a consideration when the GNA legislation was established and DoD policy for JDAL was set. Frequently, personnel that staff JTFs are in positions on service staffs or on service component staffs and are serving in a JTF as an individual augmentation. Exceptions exist, but, in general, personnel in these positions do not receive joint qualifications, either because the position does not qualify or the duration of the assignment is limited.

Cumulative Joint Duty Credit

The 1996 NDAA authorized that credit for a full JDA or credit countable for determining cumulative service[19] is awarded to officers serving in qualifying temporary JTF assignments. Cumulative credit may be earned through either of two methods:

- For service performed in a JDA that totals less than two years (general and flag officers) and less than three years (other officers) and includes at least one tour of duty in a joint assignment that was either performed outside the continental U.S. or terminated due to a reassignment[20]
- For service in Combined Joint Task Force Headquarters (CJTF) assignments in approved operations.

---

[18] COS officers can receive up to two months constructive credit if they depart their JDA for "military necessity."

[19] JDA credit awarded for certain task force assignments is exempt from JDA promotion reports, minimum tour-length requirements and military service tour-length averages, assignment fill rates and professional joint education sequencing requirements (DoDI 1300.20, 1996).

[20] A reassignment must be either for a personal reason beyond the officer's control or because an officer was promoted to a grade in which a joint duty assignment was not available to the officer within a command at his or her promoted grade, or the billet was eliminated.

Cumulative joint duty credit for continuous service performed in a JDA is granted for tour lengths that are shorter than the time needed to qualify for full joint duty credit, provided they last a minimum of ten months. The Secretary of Defense retains waiver authority for tour lengths of joint duty assignments.

The award of cumulative credit for JTF headquarters (HQ) duty in ongoing operations is currently authorized for positions in CJTF-AFG (Afghanistan), CJTF-HOA (Horn of Africa), CJTF-7 (Iraq), and other JTFs, although there are some restrictions pertaining to the award of CJTF cumulative credit.

Organization of This Report

Chapter Two provides a brief overview of the survey methodology, implementation of the survey, and survey response rates. Chapter Three describes the survey respondents in terms of some basic characteristics largely related to their military service. These characteristics include but are not limited to paygrade, service, incumbency status of the respondent, length of time in the current billet, occupation (using the Department of Defense single-digit occupational code), prior education and training, whether the respondent is a joint specialty officer (JSO), and type and location of the billet.

Chapters Four, Five, and Six focus on a particular topic and analyze the variation in responses across billet organizations/services and subordinate organizations, paygrade, and other variables of interest. Of particular interest is how these billets rank on selected indicators of "jointness," which are shown in Table 1.1. Chapter Four looks at the variety of agencies and types of personnel with whom respondents report interacting and the frequency of those interactions. Chapter Five examines the types of tasks typically performed by respondents, the time spent on those tasks, the relative (subjective) level of importance of the tasks to respondents jobs, and the level of responsibility respondents hold for each task performed. Chapter Six addresses the types of knowledge that respondents believe are required or helpful for a person in their position and whether they expect to have acquired such knowledge by the end of their assignment. It also presents data on the kinds of experience, education, and training respondents believe would be desirable or helpful for an individual to perform the duties of a position successfully and the kinds of joint experience (multiservice, interagency, multinational) the billet currently provides them.

Chapter Seven provides interesting and useful information on topics that do not fit neatly into the other chapters, including respondents' perceptions of the optimal length of assignments, the uniqueness of their experience, the ability of civilians or officers from another service to perform certain duties, and the value of temporary JTF assignments.

Chapter Eight concludes by presenting a simple illustration of how these data might be used to examine issues concerning the demand for and supply of joint duty billets, which are the focus of this current work and the main reasons for conducting the survey and collecting the data.

Appendix A contains a copy of the JOM survey form. Two survey protocols were used—one for billet incumbents and one for non-incumbents (e.g., supervisors). In this report, we include only the incumbents' survey. Appendix B presents frequencies, means, and standard deviations for the survey data and the number of missing responses in the survey sample for each survey question. Appendix C provides supporting data for the findings presented in Chapter Six.

Table 1.1
Definitions of Indicators Used to Characterize "Jointness"

Metric	Indicator
Tasks performed during the typical workweek	Percentage of officers providing strategic direction and integration Percentage of officers developing/assessing joint policies Percentage of officers developing/assessing joint doctrine Percentage of officers fostering multinational, interagency, or regional relations Percentage of officers performing three or more of these tasks
Interactions with non–own-service organizations and personnel	Median number of non–own-service organizations with whom officers interact monthly or more frequently Median number of non–own-service personnel with whom officers interact monthly or more frequently
Supervision of billet by non–own-service personnel/civilians	Percentage of officers reporting being supervised by one or more non–own-service supervisor/civilian/non–U.S. military personnel or civilian
Need for joint professional education or prior joint experience	Percentage of officers reporting that JPME II is required or desired for the assignment Percentage of officers reporting that prior joint experience is required or desired for the assignment
Types of joint experience provided by the billet	Percentage of officers reporting getting significant experience in multiservice matters Percentage of officers reporting getting significant experience in multinational matters Percentage of officers reporting getting significant experience in interagency matters Percentage of officers reporting getting significant experience in all three areas

CHAPTER TWO

Survey Methodology

This chapter presents an overview of the Joint Officer Management Census survey methodology: how billets were identified for inclusion in the survey, development of the protocol, and fielding, along with overall survey response rates. The protocol was largely developed by RAND and subsequently revised by the OSD and the Joint Staff; fielding was done by LMI, a nonprofit consulting firm that frequently works with OSD, under the guidance of this study's sponsor. This section draws heavily on conversations and materials provided by the sponsor and LMI.

Billet Identification

As mentioned earlier, the number of billets eligible for the survey was left intentionally broad and included three types of JDAL billet categories:

- Billets currently on the JDAL
- Non-JDAL billets in external organizations that have some billets on the JDAL
- Non-JDAL internal service billets nominated by the services.

The purpose of this categorization was to evaluate and identify billets that *require* joint education, experience, and/or training and those that *provide* joint experience. The services and external organizations were asked to identify: (1) billets for which a prerequisite joint education and experience tour would better qualify an officer to perform the mission requirements of the position and (2) billets that provide officers with significant experience in joint matters (for example, billets that provide incumbents with multinational, multiservice, or interagency experience) and thus could be deemed to be contributing to joint duty qualifications.

Each service used its own criteria for nominating billets. The Army and the Air Force nominated significantly more billets than the other services, and the number of service-nominated billets overall was much larger than expected. However, this was due in part to the services nominating some external billets already included in the two types of bullets mentioned above. After duplicates were excluded, the survey population consisted of approximately 30,000 billets (including JDAL billets, non-JDAL billets in external organizations, and service-nominated billets).

In fielding the survey, it was discovered that the list that formed the survey population frame was outdated and incomplete. It did not fully reflect current organizational changes and the restructuring of some organizations. As a result, some billets identified under the old systems no longer existed, and others were not reflected in the sampling frame. For example, we were informed that intelligence billets were significantly undercounted. In addition, the North Atlantic Treaty Organization (NATO) had recently reorganized billets but did not maintain a crosswalk between the old and new billets; as a result, some NATO billets may not have been included in the survey population. Moreover, some organizations were inadvertently excluded from the survey.

As a result, although the intention had been to survey all billets that met the criteria above—in other words, conduct a census of actual and potential JDA billets—the survey actually encompassed a subset of billets rather than the census (the complete universe of JDA billets). The 30,043 billets that were surveyed included 8,475 JDAL billets; 6,384 billets in external organizations (which encompassed almost all the billets in most of the organizations, with some notable exceptions in intelligence organizations); and 15,184 service-nominated billets (which presumably covered all the billets the services designated as meeting their criteria). The number of JDAL billets changes from year to year as some billets are dropped and others are added. Over the past several years, the number has been between 9,100 and 9,300 billets. In 2004, it increased to 9,700. Thus, the sampling frame encompassed 87 percent of the JDAL billets.

Table 2.1 presents the billet paygrade distribution for the billets that were surveyed. Later in this chapter, we present data on who actually responded to the survey. The paygrade distribution across the various billet groups differs quite a bit. JDAL billets are primarily at the O-4 and O-5 grades (83 percent), and another 17 percent are at the O-6 level. In contrast, with the exception of the Marine Corps, 33–43 percent of billets in the non-JDAL groups are junior-grade billets (primarily at the O-3 grade) compared with none of the JDAL billets being junior grade. About 3–5 percent of billets in the external organizations and the three services are at the general/flag officer level (O-7–O-10), compared with only 0.2 percent of JDAL billets. The Army nominated the highest proportion of junior paygrade billets—43 percent, compared with 38 percent of Navy billets and 33 percent of Air Force billets. Conversely, the Army had the smallest proportion of O-6 billets (6 percent) compared with approximately 10 to 12 percent in the other three services. The distribution of the Marine Corps billeted paygrade looks very different from that of the other services—the Corps nominated a very small number of billets (167), nearly half of which were at the general/flag officer level.

Table 2.1
Distribution of Billets by Billet Paygrade and JDAL Billet Category

			Service-Nominated Billets			
Billet Paygrade	**JDAL Billets**	**Non-JDAL Billets in External Organizations**	**Army**	**Navy**	**Air Force**	**Marine Corps**
O-1–O-3	0.0	36.3	42.9	38.1	33.2	4.2
O-4	41.7	26.7	31.1	21.6	29.6	17.4
O-5	40.8	23.2	15.5	20.4	22.9	21.6
O-6	17.3	9.7	6.2	11.9	10.9	9.6
O-7–O-10	0.2	4.1	3.8	4.7	3.3	47.3
Unknown	0.0	0.0	0.6	3.3	0.0	0.0
Total	8,475	6,384	6,085	2,270	6,662	167

Because there is considerable interest in the types of billets nominated by the services that are potentially joint billets, we present data on the actual number of billets nominated by the services grouped by service, subordinate organization, and billet paygrade (see Table 2.2). In terms of the overall distribution by subordinate organizations, of the 6,085 billets nominated by the Army, 31 percent were in Forces Command, 29 percent were in Special Operations Command, and 12 percent were in Training and Doctrine Command (TRADOC). About two-thirds of Special Operations Command billets were junior grade billets, compared with 40 percent of TRADOC billets and 25 percent of Forces Command billets. More than half of the 53 billets in the Army Materiel Command (AMC) were at the senior level (O-6 and above).

The Navy nominated 2,270 billets, of which one-third were in Net Command and 16 percent were in Pacific Fleet. Security Group and Space Warfare Command (SPAWAR) each accounted for 11 percent of billets. Net Command had the largest percentage of junior grade billets (67 percent). Half of the Naval Staff billets were at the senior level (O-6 and above).

Among the 6,662 billets nominated by the Air Force, four subordinate organizations accounted for more than 70 percent of the billets: Air Combat Command (24 percent); Air Force Elements (ELM) (19 percent); Air Force Materiel Command (16 percent); and Space Command (SPACECOM) (12 percent). Between 38 and 42 percent of the billets in these subordinate organizations were junior-grade billets, with the exception of ELM (21 percent). About half of ELM billets were at the O-5 and O-6 grades.

Because the number of Marine Corps billets was so small, there was no clear way to group them into subordinate organizations. As mentioned above, well over half were at the O-6 and above level.

Protocol

The survey protocol, which is shown in Appendix A, was developed primarily by RAND, although some modifications were added by the sponsor and LMI. The major topics covered in the survey include:

- Characteristics of the person serving in a billet: paygrade; service and component; time assigned to the billet, time spent on temporary duty travel/temporary additional duty (TDY/TAD) due to training or education during the assignment, time spent on TDY/TAD for other purposes during the assignment; education and training of the respondent; joint specialty officer status; credit for JPME I and JPME II; gender, and race/ethnicity
- Overall billet characteristics: temporary JTF billet; JTF headquarters/subordinate organization/service component; location and receipt of special pays
- Specific billet characteristics:
  - Tasks typically performed and percentage of time spent on these tasks; relative importance of and level of responsibility for these tasks
  - Interactions with different agencies, services, and types of personnel and frequency of these interactions

Table 2.2
Number of Billets Nominated by the Services, by Service, Subordinate Organization, and Billet Paygrade

Subordinate Organizations	O-1–O-3	O-4	O-5	O-6	O-7–O-10	Number of Billets	Percentage of Total
Army							
AMC	1	14	11	3	24	53	0.9
Intelligence and Security Command	110	131	21	9	1	272	4.5
Special Operations Command	1,157	441	101	42	4	1,745	28.7
Criminal Investigation Command	20	9	6	1	2	38	0.6
Corps of Engineers	1	1	1	0	8	11	0.2
U.S. Army Europe	163	185	108	10	18	484	8.0
Forces Command	763	665	363	64	49	1,904	31.3
8th U.S. Army	109	117	67	16	12	321	5.3
Military District of Washington	13	4	6	8	1	32	0.5
U.S. Army Pacific	84	49	34	17	6	190	3.1
Space and Missile Command	4	7	2	1	2	16	0.3
TRADOC	174	211	129	161	34	709	11.7
HQ	0	16	35	27	30	108	1.8
Army, Other	9	45	57	18	33	196	3.3
Total	2,608	1,895	941	377	230	6,085[a]	100.0
Navy							
Naval Staff	2	23	24	28	36	128	5.6
Naval Intelligence	11	16	6	4	6	44	1.9
Naval Systems Command	15	56	78	66	7	223	9.8
SPAWAR	65	83	68	34	1	251	11.1
Atlantic and Europe Fleet	19	44	40	29	21	161	7.1
Security Group	175	51	19	7	1	253	11.1
Pacific Fleet	64	100	128	48	13	363	16.0
Net Command	505	104	91	45	4	750	33.0
Navy, Other	8	13	10	10	17	97	4.3
Total	864	490	464	271	106	2,270[a]	100.0
Air Force							
Air Forces, Europe (AFE)	33	35	14	17	19	118	1.8
Air Training Command (ATC)	166	96	56	41	17	376	5.7
Air Force Reserve (AFR)	2	2	3	13	0	20	0.3
Headquarters Air Force (HQ HAF)	12	48	88	21	47	216	3.3
Pacific Air Forces (PAF)	93	85	33	30	14	255	3.9
Electronic Security Command	84	31	13	4	3	135	2.0
Air Forces Special Operations Command	29	140	99	22	3	293	4.4
Air Combat Command	663	537	257	128	31	1,616	24.5
Air Mobility Command	77	100	40	25	19	261	4.0
Air Force Materiel Command	452	246	216	141	27	1082	16.4
Air Force Space Command	306	222	177	90	13	808	12.3
Air Force Elements (Other)	265	377	449	160	4	1,255	19.0
Air Force, Other	26	35	49	23	21	154	2.3
Total	2,213	1,970	1,528	728	223	6,662	100.0

[a] A small number of billets (34 in the Army and 75 in the Navy) were missing billet paygrade information.

  - Types of knowledge required or helpful in the position and the level of expertise with respect to the various types of knowledge expected by the end of the tour.
- Perceptions of the person serving in billet regarding:
  - Whether the experience is different from that of his or her predecessors and whether future incumbents are likely to have the same experience
  - Whether the billet provides significant experience in multiservice, multinational, and/or interagency matters
  - The usefulness of JPME I and JPME II, and other joint training or education; prior experience in a joint environment; primary specialty; knowledge of service's capabilities in carrying out the duties of the position successfully
  - The most important knowledge/expertise needed for the assignment (for both the incumbent and his or her successor)
  - Whether the duties could be effectively undertaken by a civilian or an officer of another service
  - The optimal length of time for typical permanent and temporary joint duty assignments
  - Morale problems if joint duty credit is awarded only for some positions in the immediate organization
  - Length of time to become comfortable in a joint environment
  - For those with experience in a temporary JTF billet, whether greater understanding of a joint environment is gained in a permanent or a temporary joint duty-assignment billet.

There were two versions of the survey instrument—one for billet incumbents and the other for supervisors or personnel assigned to complete the survey. The two surveys were similar in form. We include only the former in Appendix A.

For a Web-based survey, particularly one as complex as the JOM census survey, the use of drop-down boxes with precoded answers is strongly encouraged by survey experts. This aspect of the survey design required with the creation of sets of appropriate and relevant optional answers to the various questions. In one case, job books were used to validate lists of tasks provided as answers to questions concerning typical tasks performed.

Questions requiring respondent-provided answers (such as the number of hours worked in a week or on a particular task) were programmed to accept mathematically appropriate answers. Participants were notified of their progress periodically throughout the survey to encourage them to continue and complete the survey.

Fielding the Survey

LMI fielded the survey in a staggered fashion to avoid overloading the system, although it was later determined that system overload was a nonissue. The survey was fielded from May 16 to June 20, 2005. The research sponsor identified approximately 50 points of contact who were asked to administer the survey by providing the targeted sample members with the access information and the link to the online survey. These points of contacts differed considerably in their perceptions of the importance and usefulness of the survey, and this variation may have had an impact on survey response rates among organizations (discussed further below).

Two "help" email addresses were available to survey respondents. Initially, technical questions were to be answered by LMI and policy-related questions were to be answered by the participant's point of contact or directed to the OSD. There were approximately 100 to 200 queries per day, although this number declined sharply over time as the survey progressed. Despite that most of the questions were policy related, respondents referred nearly all the questions to LMI, which then took on the responsibility of answering both types of questions within 24 hours.

The sponsor and LMI reported that several problems arose during the fielding of the survey—technical problems, an outdated sampling frame (discussed earlier), and less-than-ideal timing for the fielding of the survey.

The technical problems included the following:

- First, problems arose because of differing operating systems and differing types of firewalls at various locations. Many organizations had difficulties using the JavaScript format and requested a SIPRNet version of the survey. These problems often caused programs to time out on some systems, such that a survey page that was intended to allow for one to four hours of completion time went offline after only five or six minutes.
- Second, because of security and firewall problems, some respondents—those in intelligence jobs and, in particular, those at the National Security Agency (NSA)—were unable to complete the survey at their desks. The sponsor pointed to this limitation as a factor in the low response rates for some organizations.
- Third, a major technical problem occurred in the second week of the survey when the system experienced a 13-hour downtime period and participants were unable to continue beyond the 14th question in a 76-question survey. Although overall response rates were not drastically affected by the system failure, some respondents expressed displeasure over the interruption.
- Fourth, the first group of participants that experienced these problems included flag officers who faced an additional inconvenience with respect to estimated completion time. During the pilot—which did not include flag officers—the survey had taken 15 minutes to complete on average. In actuality, flag officers took 45 minutes to complete the survey, because they could not skip as many questions as the other officers.

In terms of timing, a three-day holiday weekend, which for many participants was a four-day weekend, fell in the middle of the survey period, and response rates appeared to decline after the break. To ameliorate any effects from the holiday break, the survey was republicized after both the system's downtime and the holiday weekend, and the survey period was extended.

Response Rates

The original intent of this research was to conduct a "census" of specified billets. However, as mentioned above, because of problems with identifying some billets and outdated information, we do not have a census.

Of the 30,043 unique billets that were identified, we received 21,214 responses—a response rate of 70.6 percent. Table 2.3 shows the response rates for the JDAL billets, non-JDAL billets in external organizations, and the service-nominated billets. About 70–72 percent of JDAL and non-JDAL external organization billets responded. Among the services, the Navy had the highest response rate, 90 percent, and the Marine Corps had the lowest, 44 percent. The Army and Air Force had response rates of 65 and 68 percent, respectively.

There was little difference in response rates by billet paygrade, although the response rate for the junior-grade billets was somewhat lower—66 percent—than the 70–73 percent response rate among the O-4–O-6 and general/flag officer billets.

Table 2.4 shows the response rates for the service-nominated billets by subordinate organization. Among Army subordinate organizations, several had response rates of 90 percent, including (among the larger ones) the 8th U.S. Army and Intelligence and Security Command. However, both Special Operations Command and TRADOC had low response rates, with less than half of the billets responding to the survey. The Navy, as seen above, had a very high response rate in general, and all the subordinate organizations had response rates of well over 80 percent with one exception—SPAWAR—but even that had a response rate of over 70 percent. Among the Air Force–nominated billets, the lowest response rates were among ELM and AFE, while PAF, the Electronic Security Command (ELC), Air Force Special Operations Command, and AMC all had response rates of 90 percent or higher. As mentioned above, less than half of those with Marine Corps billets responded, and the response rate was particularly low for general officers. Of the 79 general officer billets nominated for the survey, only 18 responded.

Table 2.5 lists the external organizations (JDAL and non-JDAL billets) that the survey covered. For the sake of completeness, we also included the four services in the table. The response rates varied considerably across organizations, ranging from a low of 16 percent for NATO to 100 percent for several agencies. The agencies with 100 percent response rates had small numbers of billets in the survey, ranging from one to 50, with two exceptions—

Table 2.3
Response Rates, by JDAL Billet Category

Billet Category	Total Number of Billets	Completed Surveys	Response Rate (%)
JDAL Billets	8,475	6,139	72.4
Non-JDAL Billets in External Organizations	6384	4,487	70.3
Service-Nominated Billets			
Army	6,085	3,953	65.0
Navy	2,270	2,046	90.1
Air Force	6,662	4,515	67.8
Marine Corps	167	74	44.3
Total	30,043	21,214	70.6

Table 2.4
Response Rates, by Subordinate Organization, Service-Nominated Billets

Subordinate Organizations	Total Number of Billets	Completed Surveys	Response Rate (%)
Army			
AMC	53	52	98.1
Intelligence and Security Command	272	245	90.1
Special Operations Command	1,745	769	44.1
Criminal Investigation Command	38	38	100.0
Corps of Engineers	11	9	81.8
U.S. Army Europe	484	424	87.6
Forces Command	1,904	1,344	70.6
8th U.S. Army	321	319	99.4
Military District of Washington, D.C.	32	18	56.3
U.S. Army Pacific	190	163	85.8
Space and Missile Command	16	14	87.5
TRADOC	709	301	42.5
HQ	108	80	74.1
Army, other	196	177	90.6
Navy			
Naval Staff	128	128	100
Naval Intelligence	44	40	90.9
Naval Systems Command	223	203	91
SPAWAR	251	177	70.5
Atlantic and Europe Fleet	161	134	83.2
Security Group	253	250	98.8
Pacific Fleet	363	304	83.7
Net Command	750	719	95.9
Navy, other	97	91	93.8
Air Force			
AFE	118	53	44.9
ATC	376	239	63.6
AFR	20	14	70
HQ (HAF)	216	158	73.1
PAF	255	249	97.6
ELC	135	126	93.3
Air Force Special Operations Command	293	278	94.9
Air Combat Command	1,616	1,088	67.3
Air Mobility Command	261	241	92.3
Air Force Materiel Command	1,082	863	79.8
Air Force Space Command	808	590	73
Air Force Elements (other) (ELM)	1,255	501	39.9
Air Force, other	154	115	74.7
Marine Corps	167	74	44.3

the Defense Threat Reduction Agency (455 billets) and the National Geospatial Intelligence Agency (229 billets). Several organizations had response rates of 90 percent or higher. NSA (at least partly as a result of the classification and accessibility issues mentioned above) had a low response rate (37 percent). Among the combatant commands, U.S. Pacific Command and the U.S. Special Operations Command had response rates of more than 90 percent, compared with a low of 58 percent for the U.S. Strategic Command.

Note that response rates are calculated as a percentage of the total number of billets that were actually surveyed. However, because of a potential undercounting of billets, the correct response rates for some organizations may actually be lower than those shown here.[1] Nonresponse and undercounting was a problem particularly in the intelligence communities, as mentioned above. In addition, the recent reorganization of some of these organizations caused new billets to be missing from the survey population frame, further adding to the problem of undercounting.

Table 2.5
Response Rates, by Billet Organization

Organization	Total Number of Billets	Completed Surveys	Response Rate (%)
Air Force	6,662	4,515	67.8
American Forces Information Service	39	38	97.4
Army	6,085	3,953	65.0
CENTCOM JTF	1,033	710	68.7
Defense Acquisition University	37	37	100.0
Defense Advanced Research Projects Agency	23	23	100.0
Defense Contract Management Agency (Combat Support Agency [CSA])	317	199	62.8
Defense Finance and Accounting Service	57	41	71.9
Defense Information Systems Agency (CSA)	476	441	92.6
Defense Intelligence Agency (CSA)	1,306	861	65.9
Defense Legal Services Agency	27	16	59.3
Defense Logistics Agency (CSA)	292	265	90.8
Defense Prisoner of War/Missing Personnel Office	30	29	96.7
Defense Security Cooperation Agency	43	43	100.0
Defense Technology Security Administration	16	13	81.3
Defense Threat Reduction Agency (CSA)	455	455	100.0
DoD Counterintelligence Field Activity	1	1	100.0
DoD Human Resources Activity	15	6	40.0
DoD Inspector General	33	33	100.0
Inter-American Defense Board (IADB)	17	17	100.0
Joint Requirements Office, Chemical, Biological, Radiological, Nuclear	11	10	90.9
Joint Theater Air And Missile Defense Organization	23	21	91.3
Marine Corps	167	74	44.3
Missile Defense Agency	117	110	94.0
National Defense University	208	202	97.1
National Geospatial-Intelligence Agency	229	229	100.0
NSA	675	248	36.7
Navy	2,270	2,046	90.1
North American Aerospace Defense Command	145	123	84.8
NATO	1,474	239	16.2
Office of Economic Adjustment	3	2	66.7
OSD	383	267	69.7

[1] Points of contact were asked to annotate and return duplicate surveys and, in addition, annotate those surveys for which no respondent was available to complete the survey. The surveys in the latter group should have been included in the total population count that is the denominator for calculating the response rates. However, some returned surveys—whether they represented duplicates or nonresponse—were mistakenly removed from the database.

Table 2.5—Continued

Organization	Total Number of Billets	Completed Surveys	Response Rate (%)
Pentagon Force Protection Agency	4	4	100.0
Joint Staff	840	812	96.7
TRICARE Management Activity	50	50	100
U.S. Central Command (CENTCOM)	734	642	87.5
U.S. European Command (EUCOM)	855	684	80.0
U.S. Joint Forces Command (JFCOM)	725	530	73.1
U.S. Northern Command (NORTHCOM)	476	394	82.8
U.S. Pacific Command (PACOM)	1,071	973	90.8
U.S. Southern Command (SOUTHCOM)	486	321	66.0
U.S. Special Operations Command (SOCOM)	629	564	89.7
U.S. Strategic Command (STRATCOM)	1,226	709	57.8
U.S. Transportation Command (TRANSCOM)	250	236	94.4
Washington Headquarters Services	28	28	100.0
Total	30,043	21,214	70.6

SOURCE: Data are calculated from the LMI master billet file and survey data.

Summary

The overall response rate for the survey was 71 percent, with a total of 21,214 completes. Response rates varied considerably across organization, and across subordinate organizations within service-nominated billets. Because of an outdated sampling frame, it is clear that the survey encompassed a large subset of billets on the JDAL, although not all of the JDAL, and additionally, may have missed some external organizations. Despite these caveats, these data offer a rich look at officers serving on joint duty assignments or assignments that appear to provide significant joint duty experience, and they provide useful insights into issues surrounding joint officer management.

CHAPTER THREE

Overview of Survey Billets and Survey Respondents

This chapter describes the survey billets and survey respondents in terms of service and component, paygrade, billet organization, occupation, education and training, and other characteristics of the billets in which respondents are serving or about which they are responding.

Because the amount of data generated by the survey is overwhelming, we sought to summarize the data in understandable and useful ways. One of the primary goals of this study is to analyze the nature of the experiences that various jobs provide and the kind of training and education those jobs require. As such, we looked for ways to classify the data that would further this goal and set the stage for analyzing and evaluating the billets. With this in mind, we grouped billets into two categories according to the nature of the billet and the organization where the billet was located:

- **JDAL Billet Category**: In this category, billets are characterized as follows:
  1. Billets currently on the JDAL
  2. Non-JDAL billets in external organizations that have some billets on the JDAL
  3. Internal service billets (which are not on the JDAL by law) nominated by the four services.
- **Major Billet Organization Category**: This category groups billets by the organization in which a billet is currently located. Organizations were classified into 15 groups. Table 3.1 provides a crosswalk between the 45 organizations shown in Table 2.5 and the 15 aggregated groups. It also provides a distribution of billets in the survey across the 15 organizations. It also should be noted that these categories are mutually exclusive—any given billet falls into only one of these groups. All service-nominated billets are grouped together by service, rather than by the location of the nominated billet; therefore, the first four organizations are the four services.

Appendix B presents data for the entire sample of 21,214 respondents. For each question on the survey, it provides the percentage of respondents falling in to the different categories or, where appropriate, the means and the standard deviations for the responses. It also shows missing data—i.e., the number of respondents who failed to answer a survey question or set of questions.

Table 3.1
Crosswalk Between Billet Organizations and Aggregated Groups in the Major Billet Organization Category

Groups in Major Billet Organization Category	Billet Organizations
Joint Staff	Joint Staff
OSD	OSD staff
CENTCOM JTF	CENTCOM JTF
Army	Army
Navy	Navy
Air Force	Air Force
Marine Corps	Marine Corps
International organizations	Inter-American Defense Board NATO
CSAs	Defense Contract Management Agency Defense Information Systems Agency Defense Intelligence Agency Defense Logistics Agency Defense Threat Reduction Agency
Other non-OSD defense agencies	Joint Theater Air And Missile Defense Organization Missile Defense Agency National Geospatial-Intelligence Agency NSA North American Aerospace Defense Command
OSD defense agencies	American Forces Information Service DoD Counterintelligence Field Activity DoD Human Resources Activity DoD Inspector General Office of Economic Adjustment Pentagon Force Protection Agency TRICARE Management Activity Washington Headquarters Services
Educational agencies	Defense Acquisition University NDU
Geographic commands	CENTCOM EUCOM NORTHCOM PACOM SOUTHCOM SOCOM
Force provider	JFCOM
Functional commands	STRATCOM TRANSCOM

Characteristics of Surveyed Billets

Table 3.2 shows some selected characteristics of the surveyed billets, which are further described in this section.

JDAL Status

As shown in the table, 28.9 percent of the billets are currently on the JDAL, while 21.2 percent are located in external organizations with billets on the JDAL. About half the billets in the responding sample had been nominated by the services.

Major Billet Organization

The Army and Air Force had nominated more than 12,000 billets for the survey, although only about 65–68 percent responded. Together, these Army and Air Force billets account for 40 percent of the responding sample, when one examines the breakout by major billet organization category. The Navy had nominated a little more than 2,200 billets and had a very high response rate of 90 percent. Overall, Navy billets account for a little less than 10 percent of the sample. The Marine Corps accounts for only 0.3 percent of the sample—that service nominated a modest 167 billets and had a relatively low response rate of 44.3 percent. The next two largest groups in the responding sample are in the combat support agencies (10.5 percent, largely dominated by the Defense Intelligence Agency [DIA], Defense Information

Table 3.2
Selected Characteristics of Billets

Characteristics	Percentage of Billets (n = 21,214)
JDAL Status	
Billets currently on JDAL	28.9
Billets in external organizations with billets on the JDAL	21.2
Billets nominated by the services	49.9
Major Billet Organization Category	
Army	18.6
Navy	9.6
Air Force	21.3
Marine Corps	0.3
Joint Staff	3.8
OSD	1.3
CENTCOM JTF	3.3
International organizations	1.2
CSAs	10.5
Other non-OSD defense agencies	4.3
OSD defense agencies	0.8
Educational agencies	1.1
Geographic commands	16.9
Force provider	2.5
Functional commands	4.5
Billet Paygrade	
O-1	0.2
O-2	2.2
O-3	22.7
O-4	32.8
O-5	26.9
O-6	11.5
O-7	1.6
O-8	1.0
O-9	0.4
O-10	0.1
Unknown	0.5

Systems Agency [DISA], and Defense Threat Reduction Agency [DTRA] billets), and billets in the Geographic Commands (17 percent).

Billet Paygrade. In terms of the billet paygrade, about 94 percent were coded at the O-3 to O-6 level and 3.1 percent at the flag officer level. We examined the match between the billet paygrade and the paygrade of the officer currently serving in that position.

Figure 3.1 shows that while the billet positions are for the most part filled by officers at the same paygrade as the billet's paygrade, there is a small level of mismatch. For example, 15 percent of billets at the O-3 level are being filled by O-4s, and 22 percent of O-4 billets are being filled by O-5s. As might be expected, the level of mismatch falls significantly the higher the level of the billet paygrade.

Billet Supervisors. One indicator of "jointness" might be whether the billet is supervised by personnel from a service other than the service of the person filling the billet, by U.S. civilians, or by foreign personnel (civilians or military officers). For incumbents from the four services, we examined the first-level and second-level supervisors and categorized them by whether they were from the same service or another service, a civilian, or non-U.S. personnel. We then created a variable that counted whether officers had zero, one, or two first- and second-level supervisors from services other than their own or who were civilians or foreign personnel.

Not unexpectedly, close to 80 percent of JDAL billets and about 75 percent of non-JDAL billets in external organizations are supervised by at least one "non–own-service" supervisor, compared with a little more than 20 percent of the service-nominated billets (see Figure 3.2).

There are some interesting differences across major billet organizations (see Figure 3.3). Officers serving OSD staff or filling OSD defense agency billets are almost all supervised by at least one non–own-service supervisor; this was true of 75 percent or more of

Figure 3.1
Distribution of Incumbents by Billet and Incumbents' Own Paygrade, O-3 Through O-6 Billets

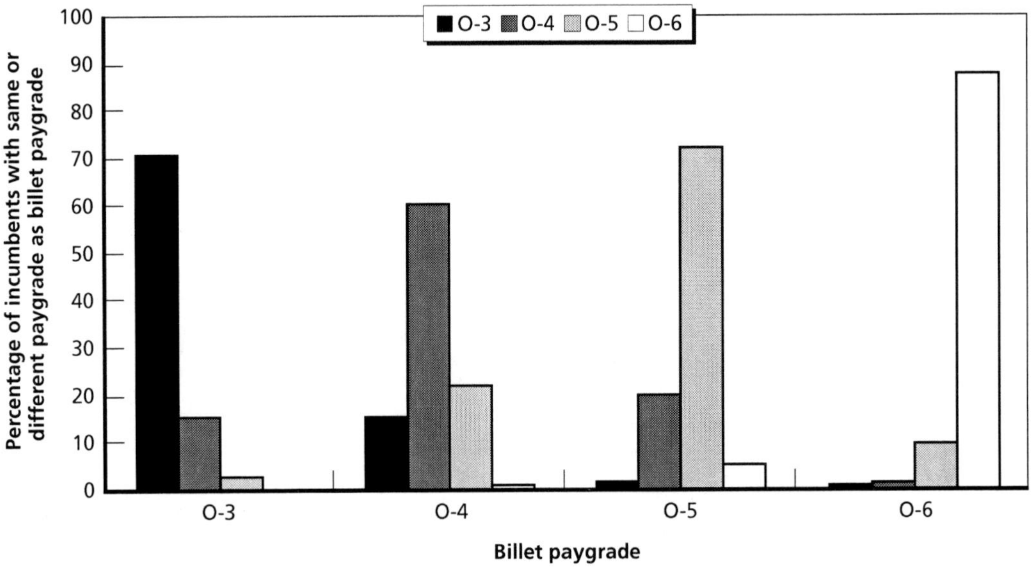

RAND *TR349-3.1*

Figure 3.2
Distribution of Officers by Number of Non–Own-Service First-Level and Second-Level Supervisors, by JDAL Billet Category

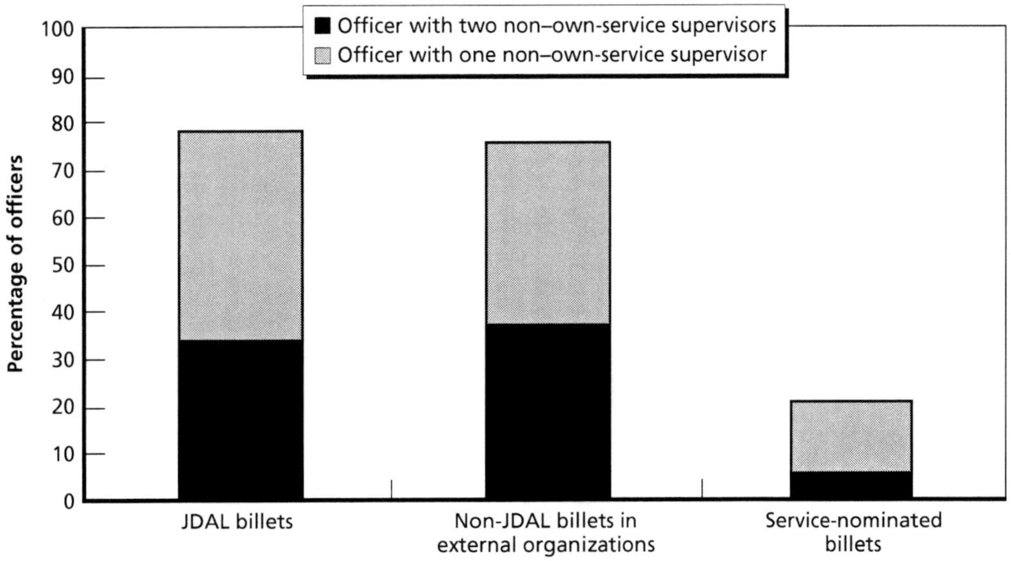

RAND *TR349-3.2*

Figure 3.3
Distribution of Officers by Number of Non–Own-Service First-Level and Second-Level Supervisors, by Major Billet Organization

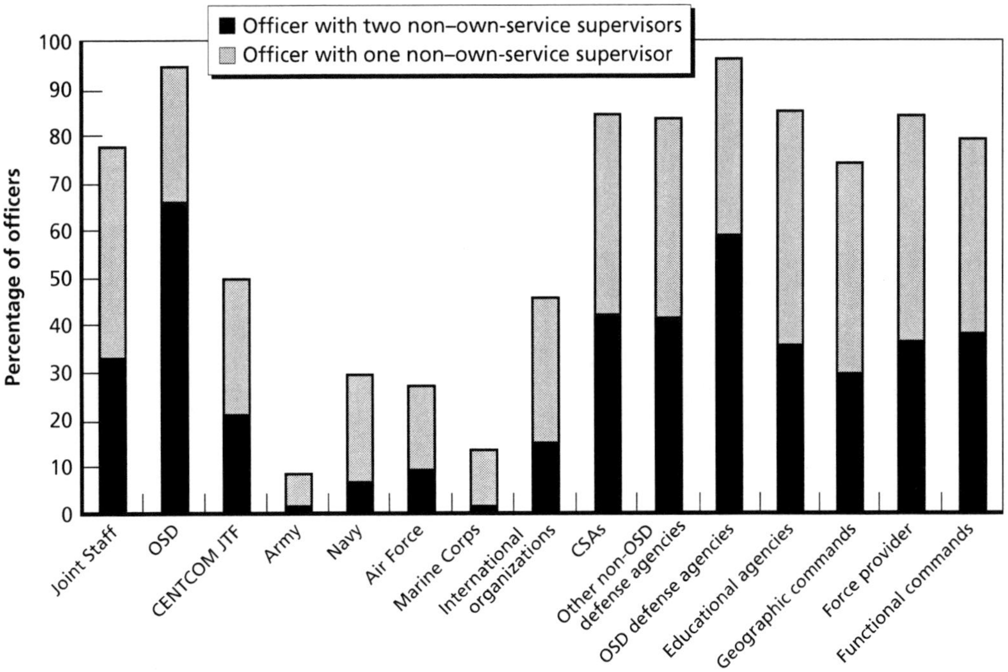

RAND *TR349-3.3*

officers assigned to other agencies, the Joint Staff, and the combatant commands. Only 45 percent of those serving in international organizations reported having at least one non–own-service supervisor. Of the major billet organizations, the four services, particularly the Army and Marine Corps, were the least likely to have supervisors from other organizations.

Type of Billet. Overall, about 61 percent of billets required serving full time with members of other military departments, and about 11 percent required serving full time with international organizations or foreign militaries. In the majority of cases, these billets remained with the originating organization or service and were not assigned to another department, organization, or foreign military.

As expected, there were marked differences by JDAL category in the types of billets, as shown in Table 3.3.[1] At least 90 percent of JDAL and non-JDAL non-service billets were involved full-time with other military departments, compared with 30 percent of service-nominated billets. About 7–10 percent of billets were specifically identified as "dual hat" positions, and most of them fell into one or both of the categories above—i.e., serving full-time with other military departments or with foreign militaries or international organizations.

Respondents were asked if they were serving in JTF billets. Overall, of the 21,214 billets, about 8 percent were assigned to JTF headquarters, another 6 percent to JTF subordinate organizations, and about 5 percent to a JTF service component.

Table 3.3
Characteristics of Billets, by JDAL Category

	Percentage of Total		
Selected Characteristics	**JDAL Billets (n = 6,139)**	**Non-JDAL Billets in External Organizations (n = 4,487)**	**Service-Nominated Billets (n = 10,588)**
Serving full-time with members of another military department	94.0	89.5	30.0
Serving full-time with members of another military department *and* assigned to a billet in another military department	11.1[a]	11.8[a]	3.4[a]
Serving full-time with armed forces of another nation or with an international military or treaty organization	15.9	11.4	8.3
Serving full-time with armed forces of another nation or with an international military or treaty organization *and* assigned to billet in a foreign military or international department	7.6[a]	4.6[a]	2.0[a]
Assigned to both own service and joint, combined, or international organization (dual-hat positions)	8.6	9.7	6.9
Joint Task Force Headquarters Staff	9.5	14.8	4.9
Joint Task Force Subordinate Organization	3.5	7.1	7.1
Joint Task Force Service Component	2.4	5.5	7.0
Permanently assigned to JTF Service Component	1.3[a]	2.7[a]	3.1[a]
Joint Program Office	3.8	8.1	4.8

NOTE: Totals may not add to 100 percent due to respondents checking more than one category in their answers.
[a]These numbers are stated as a percentage of the total number of billets. Thus, they are included in the percentages shown in the first line of that category.

[1] There appeared to be some confusion among survey respondents over the wording of the questions in Table 3.3 on the types of billets in which individuals served; therefore, the data presented in this subsection need to be interpreted cautiously.

There are some differences by JDAL status among the billets. For example, about 15 percent of those in non-JDAL billets in external organizations were serving at JTF headquarters, compared with 10 percent of those assigned to JDAL billets and 5 percent of those assigned to service-nominated billets. Compared with JDAL billets, higher proportions of non-JDAL billets were at a JTF subordinate organization or JTF service component. Of those in the JTF service components, about half were permanently assigned to the JTF. Between 4 and 8 percent of the total served in a Joint Program Office.

There was little overlap among these responses. About 2.5 percent of respondents reported serving in JTF headquarters staff and in a JTF subordinate organization, and 3 percent reported serving in a JTF subordinate organization and in a JTF service component.

When asked specifically whether the billets being evaluated were temporary JTF billets, less than 5 percent (1,026) of respondents answered in the affirmative. These billets were largely Army, Air Force, and Navy-nominated billets (29, 17, and 10 percent, respectively). About 26 percent were CENTCOM JTF billets, and 7 percent were located in the Geographic Commands.

Billet Location and Eligibility for Various Pays and Allowances. Table 3.4 shows the geographic location of the billets and the types of pays and allowances for which the billets are eligible. About four-fifths of all billets in the sample are in the United States and 7.6 percent are in Europe. Iraq and the Middle East account for about 4 percent of billets, and Korea accounts for 3.5 percent. About 10 percent of billets overall are eligible for various pays when the individual is located away from family or serving under a hardship or hostile conditions.

Characteristics of All Survey Respondents[2]

Incumbency Status

Table 3.5 shows the incumbency status of the survey respondents. Of the 21,214 respondents, about 81 percent were billet incumbents, 6 percent were supervisors of the billet, and the remaining 13 percent were other individuals designated by the points of contact to complete the survey on behalf of the incumbent.

Service Affiliation and Paygrade

Table 3.6 shows the service affiliation and paygrade of survey respondents, by the three JDAL billet categories.

Service Affiliation. Among JDAL billet respondents, about 34 percent were Army and Air Force officers, 22 percent were Navy officers, and 6 percent were Marine Corps officers. Air Force officers were the largest group among those serving in non-JDAL billets in external organizations (34 percent), while Army and Navy officers accounted for 24 and 20 percent, respectively. Among those billets nominated by the services, about 43 percent were occupied by Air Force officers, 37 percent by Army officers, 20 percent by Navy officers, and

[2] Percentages shown in this section are for nonmissing cases. The number of missing cases for each question is provided in Appendix B.

Table 3.4
Location of Billet and Eligibility for Pays and Allowances

Selected Characteristics	Percentage (n = 21,214)
Location	
United States (including Alaska and Hawaii)	79.9
Iraq	2.4
Other Middle East	1.8
South Asia (e.g., Pakistan, Afghanistan)	1.1
Korea	3.4
Cuba	0.0
Europe	7.6
Other nations outside the United States	3.5
Afloat at sea	0.3
Eligibility for pays and allowances	
Family Separation Allowance	10.7
Hostile Fire or Imminent Danger Pay	10.5
Hardship Duty Pay	8.0
Combat Zone Tax Exclusion	10.2

Table 3.5
Incumbency Status of Survey Respondents

	Percentage (n = 21,214)
Respondent was:	
Billet incumbent	80.8
Supervisor of the billet	6.1
Another person designated to complete the survey	13.1

Table 3.6
Selected Characteristics of Respondents, by JDAL Billet Category

Selected Characteristics	JDAL Billets (n = 6,139)	Non-JDAL Billets in External Organizations (n = 4,487)	Service-Nominated Billets (n = 10,588)
Service Branch and Component			
U.S. Army	33.1	23.6	36.0
U.S. Army Reserve	0.9	6.3	0.5
U.S. Army National Guard	0.1	1.5	0.1
U.S. Navy	21.3	19.5	18.6
U.S. Naval Reserve	0.7	1.9	0.9
U.S. Air Force	34.8	34.2	42.4
U.S. Air Force Reserve	0.3	1.4	0.5
U.S. Air National Guard	0.1	0.3	0.1
U.S. Marine Corps	6.2	4.7	0.6
U.S. Marine Corps Reserve	0.1	1.7	0.1
Not applicable (proxy cases only)	2.5	4.8	0.2
Paygrade			
O-1–O-3	4.0	26.8	35.9
O-4	30.8	26.6	26.8
O-5	43.9	25.6	21.4
O-6	17.3	11.2	9.1
O-7–O-10	0.5	3.0	5.4
Civilian: GS-11–GS-15[a]	3.5	6.9	1.3

[a] Nonincumbents.

less than 1 percent by Marine Corps officers. Another 3–5 percent of JDAL and non-JDAL billets in external organizations reported being "not applicable"; these were all proxies who were civilians.

Among the incumbents, 679 officers identified themselves as reserve officers. Because the sample sizes are so small, we do not report separate breakouts by active/reserve status in the remainder of this report, although we do recognize that the experiences and perceptions of the two groups of officers may differ.[3]

Paygrade Distribution. The grade distribution differs significantly by JDAL status of the billets. While only 4 percent of those assigned to JDAL billets were junior officers, the percentage was considerably higher in the other two groups: 27 percent and 36 percent. This result is not surprising, given that the services and external organizations were asked to nominate billets that provided or required joint experience or training and to specifically include junior-grade billets. In terms of overall grade distribution, O-4s to O-6s accounted for close to 70 percent of the survey respondents; a little less than a quarter were O-1s to O-3s; 3.5 percent were flag officers; and about 3.1 percent were civilians.

Characteristics of Billet Incumbents

For some analyses, it makes more sense to examine the characteristics of incumbents only.

Length of Service

On average, incumbents had served 16 years as commissioned officers. About a quarter had ten years of service or less, while the top quartile had 21 or more years of service.

Time in Billet

Incumbents had spent about 16 months on average in the billets surveyed. The median time spent on an assignment was one year, and one-quarter of respondents had spent between 23 and 48 months in a billet. Incumbents also reported spending an average of 9.5 weeks on TDY/TAD (not including training or education) during this assignment and close to three weeks on TDY/TAD due to training or education. It should be noted that close to half of the incumbents had not spent time on training or education during their incumbency.

Occupational Distribution

In terms of the distribution of officers across occupational areas using the one-digit DoD occupational code, close to 38 percent of incumbents were working in jobs classified as part of tactical operations. About 20 percent were supply and procurement officers, 12 percent were intelligence officers, and another 12 percent were engineering/maintenance officers. About 10 percent identified themselves as scientists and professionals or health care officers, while 6 percent served as administrators. About 1.3 percent were general officers.

Figure 3.4 shows the distribution of officers by occupation for the three JDAL billet categories. Much higher proportions of officers assigned to JDAL or service-nominated

[3] This is the subject of a future RAND report in development as of this writing.

Figure 3.4
Distribution of Officers by Occupation and JDAL Billet Status

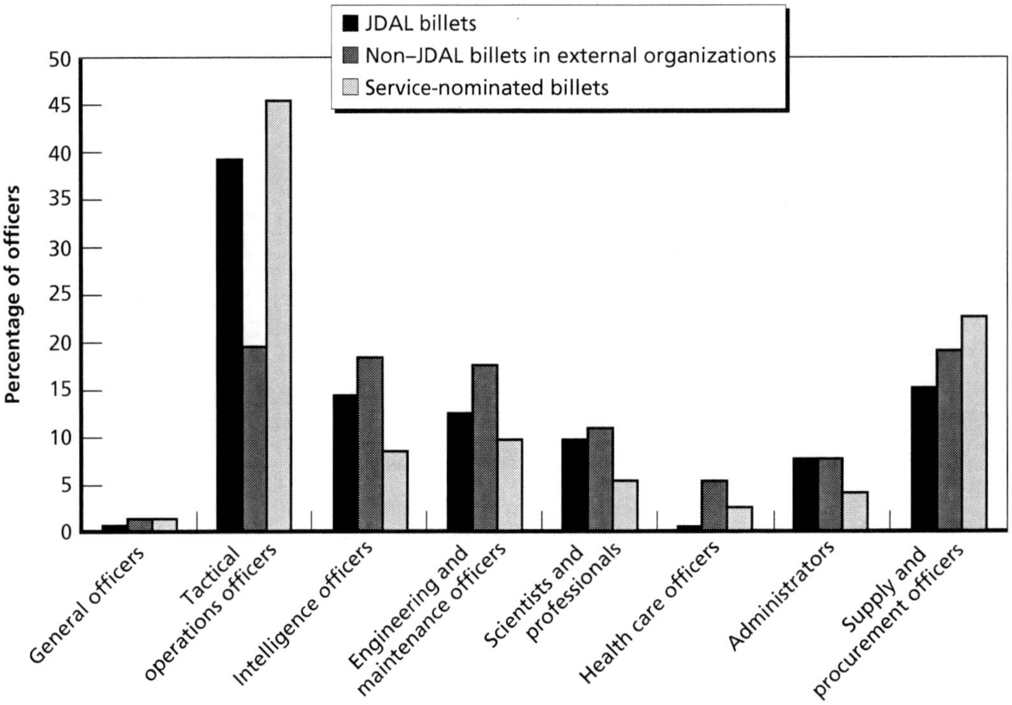

RAND *TR349-3.4*

billets were tactical operations officers compared with those in non-JDAL billets in external organizations. More engineering and maintenance and intelligence officers served non-JDAL billets than in the other two categories of billets.

Education, Credit for JPME I and JPME II, and JSO Status of Incumbents, Grades O-4 and Higher

This section examines the training and education received by incumbent officers who participated in the survey and whether they received credit for JPME I or JPME II. We also examine JSOs and their distribution across the various organizations. This analysis is limited to officers who are O-4 or higher in rank.

It might be helpful to repeat the definition of Joint Professional Military Education from Chapter One. JPME focuses specifically on joint matters. An officer can get JPME I at intermediate-level and senior-level colleges (resident or nonresident) or at NDU. JPME II is the joint professional education required prior to selection as a JSO; it is offered at the JFSC and the JCWS at the JFSC. The NWC and ICAF provide full JPME I and JPME II credit for graduates.

Thus, to meet the educational prerequisites to become a JSO/JSO nominee, officers currently must complete one of the following:

- JPME I at (an accredited) service intermediate-level or senior-level college, followed by JPME II at JFSC.[4]

---

[4] Officers (other than those with a critical occupational specialty) must attend JPME II prior to completing their joint assignment to qualify as a JSO. Attendance at JPME II prior to completing JPME I requires a waiver by CJCS.

- An intermediate-level or senior-level international military education program for which JPME I credit has been approved by CJCS, followed by JPME II at JFSC.
- The course of instruction at either NWC or ICAF.

The most common career path to becoming a JSO is to attend JPME II first, serve a joint duty tour as a JSO Nom (officers nominated to become JSOs), and then receive the JSO designation. With one exception, JPME II is a prerequisite for JOS/JSO Noms.[5]

Attendance at Intermediate- and Senior-Level Schools. Table 3.7 shows attendance at the various intermediate- and senior-level schools by service affiliation of the incumbent for officers who are O-4 or higher in rank.

Among incumbents, Navy officers generally had the highest attendance rates at joint and combined schools. While 15–20 percent of Army, Air Force, and Marine Corps officers

Table 3.7
Attendance at Intermediate and Senior Schools, by Service Affiliation of Incumbent Respondents, O-4 and Above

Intermediate/Senior Schools	Percentage of Responding Officers			
	Army	Navy	Air Force	Marine Corps
Joint and Combined Schools				
National War College	5.8	11.6	6.5	8.3
Industrial College of the Armed Forces	3.1	6.2	3.6	4.2
School of Information Warfare and Strategy	0.3	0.8	0.4	0.2
Joint Forces Staff College	15.4	31.0	19.6	15.9
Joint and Combined Staff Officer School	4.3	2.8	2.2	1.8
Joint and Combined Warfighting School	2.5	2.6	1.8	3.4
Army Service Schools				
U.S. Army War College (resident)	10.9	1.7	1.1	2.7
US Army War College (nonresident)/Department of Distance Education	2.7	0.0	0.1	0.5
Army Command and General Staff College (resident)	54.6	6.3	4.1	5.8
Army Command and General Staff College (nonresident)	39.7	0.1	0.2	0.5
Navy Service Schools				
College of Naval Warfare	1.1	13.8	1.7	5.1
College of Naval Command and Staff (resident)	2.8	23.2	1.7	7.8
College of Distance Education (Navy intermediate-level college nonresident)	0.1	12.4	0.3	4.0
Air Force Service Schools				
Air War College (resident)	0.5	1.7	7.2	1.3
Air War College (nonresident)	0.7	2.5	35.3	3.8
Air Command and Staff College (resident)	2.2	5.5	23.2	1.3
Air Command and Staff College (nonresident)	0.5	11.0	79.1	1.3
Marine Corps Service Schools				
Marine Corps War College	0.1	0.5	0.2	2.5
Marine Corps Command and Staff College	0.9	4.4	1.9	38.6
Marine Corps College of Continuing Education (Marine Corps Command and Staff College nonresident)	0.4	0.1	1.6	57.1

[5] See Chapter Four of Harrell et al. (1996) for a succinct description of how JSOs are produced and the issues related to producing sufficient JSOs and JSO Noms to fill critical and noncritical JDAs.

had attended JFSC, a much higher percentage—31 percent—of Navy officers had done so. About 6–8 percent of the Army, Air Force, and Marine Corps officers had graduated from NWC, compared with 12 percent of Navy officers. Only between 2 and 3 percent had attended JCWS. Between 3 and 6 percent had graduated from ICAF, and the number was again highest among Navy officers.

Not unexpectedly, officers tended to attend their own service schools. For example, 55 percent of Army officers had completed Army Command and General Staff College (resident), 35 percent of Air Force officers had attended the Air War College (resident), and 39 percent of Marine Corps officers had attended the Marine Corps Command and Staff College. Navy officers overall had the lowest attendance rates at their own service schools, with 23 percent attending the College of Naval Command and Staff (resident); this figure was even lower than the 31 percent of Navy officers who reported attending JFSC. There was a substantial difference among the services in terms of relative attendance at resident and nonresident programs: The majority of incumbent Air Force and Marine Corps officers (79 percent and 57 percent, respectively) completed nonresident programs, while only 40 percent of Army officers and 12 percent of Navy officers reported doing so.

Credit for JPME I and JPME II, JSO Status. Across all incumbent officers who were O-4 and higher, 52 percent reported that they had received credit for JPME I, while 12 percent were unsure of whether they had received credit. About 23 percent had received credit for JPME II, and 10 percent were unsure of whether they received credit. A little more than two-fifths (43 percent) of those who had received credit for JPME I had received credit for JPME II.

Almost all the officers who attended JFSC reported receiving credit for JPME II, as did 70–77 percent of those attending ICAF and JCWS.[6] In contrast, only 44 percent of officers who attended NWC claimed such credit. About 11–14 percent of officers attending ICAF and NWC seemed uncertain of whether they had received such credit.

Not surprisingly, more-senior officers were much more likely than officers in lower grades to report having received credit for JPME II. For example, 42 percent of O-6s and 40 percent of general and flag officers had received credit for JPME II, compared with 24 percent of O-5s and 10 percent of O-4s.

About 14 percent of officers in grade O-4 and above identified themselves as JSOs, but another 10 percent were unsure of whether they were JSOs.[7] Most JSOs were senior officers. For example, 44 percent of general and flag officers and 32 percent of O-6s were JSOs, compared with 12 percent of O-5s and 3 percent of O-4s.[8] The overwhelming majority (84 percent) of JSOs had received credit for both JPME I and JPME II. A small percentage

[6] Credit for JPME I and II is automatic, so it is surprising that these officers were unsure of whether they had received credit for joint education. While we could assign such credit to officers reporting that they attended the accredited schools, we chose not to do so here because we are simply reporting survey results that are all based on self-reports. We believe that this level of uncertainty and confusion is worth noting. In follow-on analytic work, we plan to examine this issue in greater detail.

[7] We could assign JSO/JSO Nom status to those who meet the criteria. Again, we chose not to do so for the reasons stated in the previous footnote.

[8] The percentage of JSOs among O-6s serving in the billets included in the survey is higher than the percentage of JSOs in the overall inventory of officers. For example, Thie et al. (2005) show that the percentage was less than 10 percent among non–health-care O-6s in the four services in 2001. The numbers shown here may indicate that JSOs are being assigned to positions where they are perceived as being most needed, even in internal service billets.

had received credit for JPME I but not for JPME II, and an equally small percentage had not gone through either JPME I or JPME II.

Figure 3.5 shows the percentage of officers (O-4 and higher) serving in JDAL and non-JDAL billets who had received credit for JPME I or JPME II and their JSO status. More than 70 percent of those assigned to JDAL billets had received credit for JPME I, about 35 percent had received credit for JPME II, and 17 percent were JSOs. In non-JDAL billets, about 40 percent had received credit for JPME I, 15 percent had received credit for JPME II, and about 10–13 percent were JSOs. There was little difference among officers serving in these non-JDAL billets, although the percentage of JSOs was somewhat higher among those serving in internal service billets. The billets nominated by the services appear to be similar to non-JDAL, non-service billets in terms of levels of education and training of officers assigned to these billets.

Another way of examining these data is to ask where officers with JPME II and JSO status are serving, although such patterns tend to be driven largely by the relative size of the group. Figure 3.6 shows the distribution of officers with these credentials across the three types of billets. While 61 percent of officers with JPME II are serving in JDAL billets, about 28 percent are serving in service-nominated billets—a percentage considerably higher than the 12 percent serving in non-JDAL, non-service billets. About 40 percent of JSOs are serving in service-nominated billets, compared with 47 percent in JDAL billets and 13 percent in non-JDAL, non-service billets.[9]

Figure 3.5
Percentage of Incumbent Officers in Each JDAL Billet Category Who Had Received Credit for JPME I or JPME II and Had JSO Status, O-4 and Higher

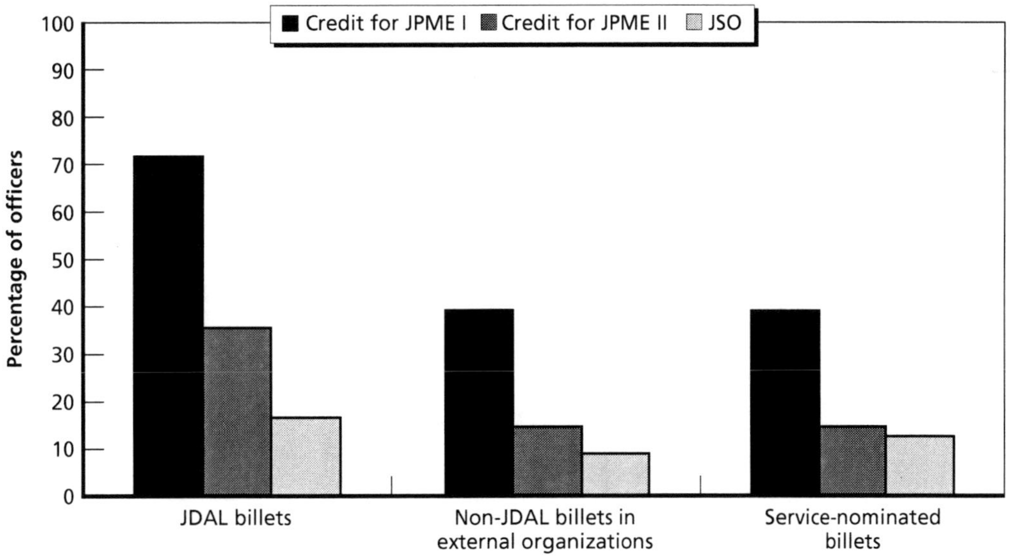

RAND *TR349-3.5*

[9] JSOs serve in large numbers in internal service billets. In follow-on research, we will examine whether this finding is due to the requirements of these positions for individuals with prior joint experience and education.

Figure 3.6
Distribution of Incumbent Officers with Credit for JPME II and JSO Status Across the Three JDAL Billet Categories, O-4 and Higher

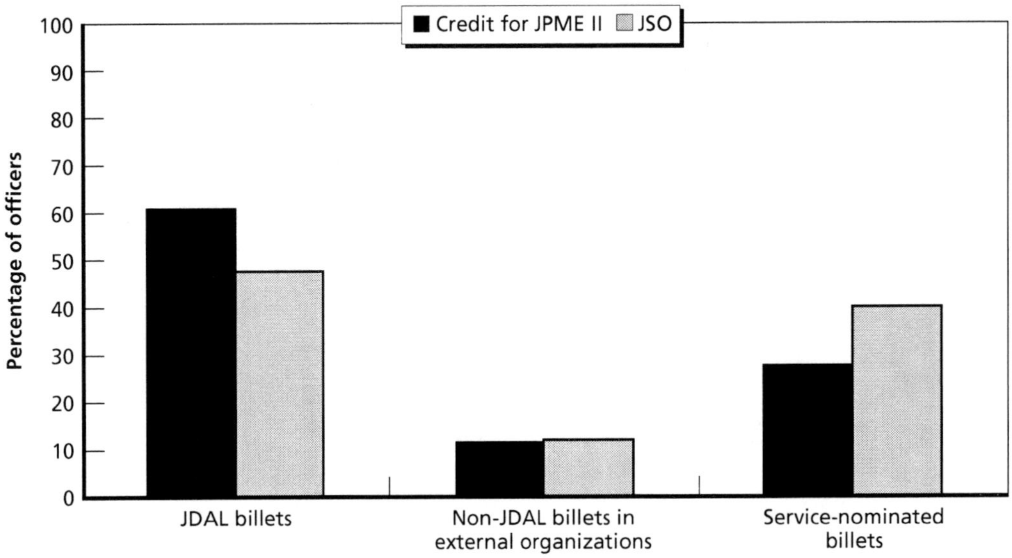

RAND *TR349-3.6*

Figure 3.7 shows the percentage of officers, O-4 and above, who received credit for JPME II and their JSO status, by major billet organization. The educational organizations rank first on both indicators. More than 60 percent of officers in educational agencies and 30–40 percent of those in force provider, international organization, functional command, OSD, Joint Staff, and geographic command billets had received credit for JPME II. Among the service-nominated billets, between 26 and 31 percent of Marine Corps officers had credit for JPME II or were JSOs, compared with 10–16 percent of officers in the other services' billets. It should be remembered that the sample size for the Marine Corps is very small, and it is likely that the Marine Corps was more selective in its nominations.

Looking at the distribution of JSOs across organizations, we find that about 22 percent of JSOs are serving in the geographic commands, 18 percent in internal Air Force billets, 12 percent in CSA billets, and 10 percent in internal Army billets.

Summary

About 29 percent of the billets in the responding sample are currently on the JDAL, and 21 percent are located in external organizations with billets on the JDAL. About half the billets in the responding sample had been nominated by the services. In contrast to the JDAL billeted grades, about 36 percent of the service-nominated billets were at the O-3 level, as were 27 percent of billets in external organizations. Overall, about 61 percent of billets required serving full time with members of other military departments, and about 11 percent were full time with international organizations or foreign militaries. About four-fifths of those answering the survey were billet incumbents. The remaining were billet supervisors or someone designated to answer the survey.

Figure 3.7
Percentage of Officers in Major Billet Organizations Who Received Credit for JPME II and Had JSO Status, O-4 and Higher

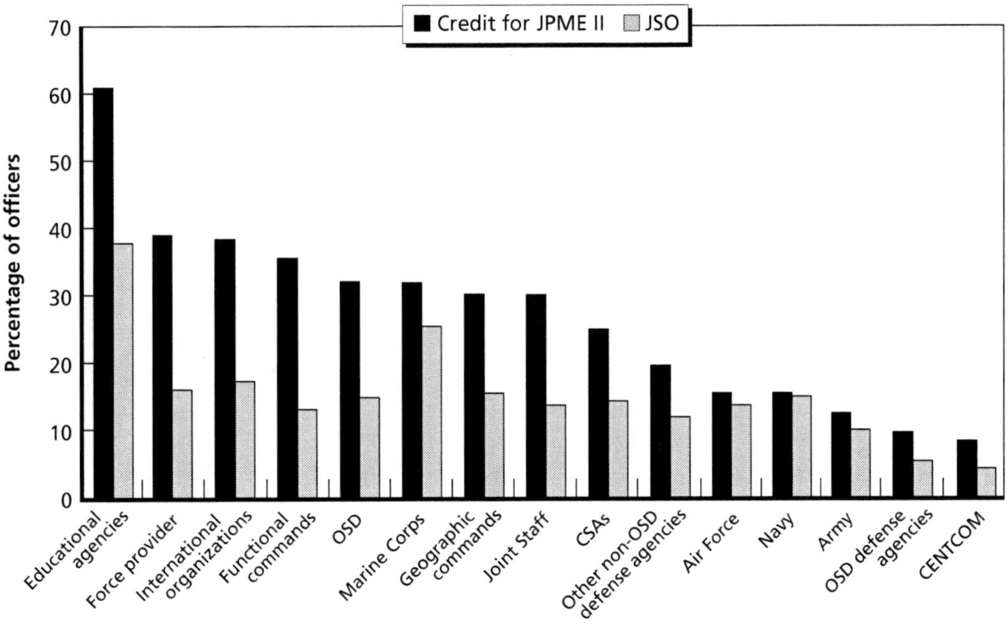

RAND *TR349-3.7*

We identify possible indicators of "jointness" and use them to rank billet categories. This sort of identification offers a useful way to summarize the data across several indicators. One caveat that should be kept in mind is that these rankings are based on data aggregated across the group, and it is entirely possible that subordinate organizations or individual organizations may rank higher or lower than the overall group. As a result, one would need to examine finer-grained data to really measure how "joint" a billet or set of billets really is. Nonetheless, we believe that the summary tables in this and subsequent chapters are useful for examining patterns across various indicators for large groupings of billets.

Here, we use percentages of officers with JPME II and who are JSOs and percentages of billets with one or more non–own-service supervisors to rank the various billet categories, as shown in Table 3.8 on the following page. A rank of 1 indicates that the organization ranks high on the indicator. Clearly, these are not the only indicators that need to be considered when examining the jointness or potential jointness of billets; subsequent chapters present other indicators of jointness.

Not surprisingly, JDAL billets rank first on all three indicators, but the service-nominated billets seem to do about as well as non-JDAL billets in external organizations. Among the major billet organizations, educational agencies rank very high, as do OSD staff and functional command billets. In contrast, CENTCOM JTF rank low on these indicators. With the exception of the Marine Corps, the services rank fairly low. In the next chapter, we turn to examining a typically used metric of jointness—the frequency and number of interactions with other organizations and types of personnel, particularly those outside one's own service.

Table 3.8
Rankings of Billet Categories Based on Selected Indicators of Education, JSO Status, and Number of Non–Own-Service Supervisors

	Rank Based on Percentage of Officers with JPME II	Rank Based on Percentage of JSOs	Rank Based on Percentage of Officers with One or More Non–Own-Service Supervisors
JDAL Category			
JDAL billets	1	1	1
Non-JDAL billets in external organizations	2[a]	3	2
Service-nominated billets	2[a]	2	3
Major Billet Organization			
Joint Staff	8	10	8
OSD	5	6	2
CENTCOM JTF	15	15	10
Army	13	13	15
Navy	12	7	12
Air Force	11	9	13
Marine Corps	6	2	14
International organizations	3	3	11
CSAs	9	8	4
Other non-OSD defense agencies	10	12	6
OSD defense agencies	14	14	1
Educational agencies	1	1	3
Geographic commands	7	5	9
Force provider	2	4	4
Functional commands	4	11	7

[a]Because the percentage difference between these two categories was only 0.1, we assigned them an equal rank.

CHAPTER FOUR

Officers' Interactions with Organizations and Various Types of Personnel

The JOM survey presented respondents with a list of 73 organizations and nine different types of personnel with whom they might interact in the course of their assignment. For those with which they interacted, respondents were asked to indicate the frequency of interaction, ranging from "less than yearly" to "daily." For purposes of this report, we focus largely on (1) *frequent interactions,* defined as monthly or more frequent interaction—i.e., "monthly," "multiple times monthly," "weekly," "multiple times weekly," or "daily"; (2) differences in the number of organizations and types of personnel with whom individuals in JDAL and non-JDAL billets interact frequently; and (3) differences in the number of organizations and types of personnel with whom individuals in major billet organizations interact frequently.

Officers' Responses on Frequency of Interaction with Organizations

While it may be important to consider the wide array of organizations with whom officers in current or potential joint duty assignments interact, there is considerable difference in the degree of "jointness" of billets when individuals interact with 30 organizations less than once a year (e.g., at a conference) and never frequently and when individuals interact with ten organizations but on a monthly or more frequent basis.

Table 4.1 illustrates this point—it shows the percentage of officers in billets on the JDAL who reported "any" and "frequent" interaction with organizations. The responses are sorted in descending order based on the responses to "frequent" interaction. Although the rank order of organizations for those reporting any interaction and rank order of organizations for those reporting frequent interaction are largely similar, there are differences between the two categories. While three-quarters of officers in JDAL billets reported any interactions with the Joint Chiefs of Staff office, only 55 percent reported that these interactions were frequent. Similarly, while 66 percent reported any interaction with CENTCOM, only 40 percent interact on a monthly or more-frequent basis. Some differences are very large (see for example, SOUTHCOM, TRANSCOM, and SPACECOM, where the difference in percentages of officers reporting any interaction versus frequent interaction is more than 30 percentage points).

More than 25 percent of respondents reported interacting frequently with 15 organizations—OSD, Joint Chiefs of Staff (JCS), the services, combatant commands, DIA, and non–U.S. military.

Table 4.1
Percentage of Officers in JDAL Billets Reporting Any Interaction or Frequent Interaction with Organizations

	Any Interaction	Monthly or More Frequent Interaction
DoD—Joint Chiefs of Staff	76.1	54.9
DoD—Office of the Secretary of Defense	71.6	46.8
DoD—U.S. Air Force	66.8	46.4
DoD—U.S. Army	68.1	45.2
DoD—U.S. Navy	63.3	41.2
DoD—CENTCOM	66.1	39.6
DoD—PACOM	64.8	35.7
DoD—EUCOM	63.3	35.2
DoD—JFCOM	59.6	30.7
DoD—U.S. Marine Corps	54.1	30.3
DoD—SOCOM	56.9	28.7
DoD—NORTHCOM	56.6	26.6
DoD—STRATCOM	55.2	25.9
DoD—Defense Intelligence Agency	49.0	25.4
Non–U.S. military	47.5	25.2
DoD—SOUTHCOM	52.3	21.0
U.S. Department of State	44.4	20.9
DoD—TRANSCOM	51.0	20.3
Central Intelligence Agency	43.1	18.6
DoD—National Security Agency—Central Security Service	37.9	14.4
DoD—National Geospatial-Intelligence Agency (formerly NIMA)	40.2	13.4
Treaty organizations (such as NATO)	35.9	11.8
DoD—Defense Threat Reduction Agency	39.1	11.0
DoD—SPACECOM	42.1	10.7
DoD—U.S. Army National Guard	40.1	10.6
DoD—Defense Information Systems Agency	33.5	9.0
DoD—U.S. Army Reserve	38.4	8.9
DoD—U.S. Coast Guard	34.5	8.4
DoD—Defense Security Cooperation Agency	31.4	7.5
DoD—U.S. Air National Guard	36.6	7.5
DoD—National Reconnaissance Office	34.7	7.5
DoD—U.S. Air Force Reserve	36.1	7.2
DoD—Missile Defense Agency	33.5	7.1
DoD—U.S. Naval Reserve	36.3	6.9
DoD—Defense Logistics Agency	33.2	6.9
Executive Branch	28.3	6.0
Other independent agency or government corporation	27.1	5.6
DoD—Defense Finance and Accounting Service	29.2	5.6
DoD—U.S. Marine Corps Reserve	34.3	5.3
DHS—Other	27.4	4.9
DoD—National Defense University	33.9	4.3
Legislative Branch	28.1	3.9
U.S. Department of Justice	26.4	3.3
DHS—Bureau of Customs and Border Protection	27.2	3.2
DoD—Department of Defense Field Activities	27.5	3.2
U.S. nongovernmental organizations (such as the American Red Cross)	27.1	3.0

Table 4.1—Continued

	Any Interaction	Monthly or More Frequent Interaction
Department of Homeland Security (DHS)—Bureau of Immigration and Customs Enforcement	26.5	2.8
DHS—Transportation Security Administration	26.2	2.5
DHS—Federal Emergency Management Agency	27.7	2.4
United Nations	30.6	2.2
DHS—U.S. Secret Service	26.1	2.2
U.S. Department of Energy	25.8	2.2
DoD—Joint Forces Staff College	32.0	2.1
DoD—Defense Advanced Research Project Agency	30.0	2.1
DoD—Defense Contract Management Agency	26.6	2.1
DoD—Industrial College of the Armed Forces	29.1	1.6
DoD—DoD Computer Emergency Response Team	25.0	1.6
U.S. Department of Commerce	23.7	1.6
U.S. Department of the Treasury	23.6	1.6
U.S. Department of Transportation	24.2	1.5
DoD—Defense Commissary Agency	25.2	1.3
Foreign nongovernmental organizations (such as the Red Crescent)	24.0	1.1
DoD—Joint Military Intelligence College	27.1	1.0
DHS—Emergency Preparedness and Response Directorate	24.4	0.9
DoD—Army Research Laboratory	27.2	0.9
DoD—Information Resource Management College	26.9	0.9
DoD—Defense Legal Services Agency	23.9	0.7
DoD—Defense Contract Audit Agency	24.9	0.7
U.S. Department of Health and Human Services	22.7	0.6
U.S. Department of Agriculture	22.7	0.5
U.S. Department of the Interior	22.5	0.4
DHS—Federal Law Enforcement Training Center	23.6	0.3
Judicial Branch	22.6	0.2

A large number of organizations (48 of 73, or 66 percent of those listed) does not rank high on the list of organizations with which respondents in JDAL billets interact frequently, with less than 10 percent of respondents reporting frequent interactions.

One of the motivations behind this survey was to understand similarities and differences among JDAL billets, non-JDAL billets in external organizations, and service-nominated billets. We selected the 15 organizations with which officers in JDAL billets interacted frequently and looked at the relative frequency of interaction among those in non-JDAL billets, as shown in Table 4.2.

By and large, officers serving in non-JDAL billets tend to have less-frequent interaction with OSD, JCS, and the combatant commands but similar levels of interaction with the services. However, service-nominated billets rank high in frequency of interaction with services but relatively low on interactions with other organizations. An interesting deviation from this pattern is the percentage reporting frequent interaction with non–U.S. military organizations—20–25 percent across the three types of billets.

Table 4.2
Percentage of Officers in JDAL Billets, Non-JDAL Billets in External Organizations, and Service-Nominated Billets Reporting Frequent Interactions with Selected Organizations

	JDAL Billets	Non-JDAL Billets in External Organizations	Service-Nominated Billets
DoD—Joint Chiefs of Staff	54.9	28.1	8.3
DoD—Office of the Secretary of Defense	46.8	31.9	13.0
DoD—U.S. Air Force	46.4	44.6	52.8
DoD—U.S. Army	45.2	45.6	38.8
DoD—U.S. Navy	41.2	41.7	32.9
DoD—CENTCOM	39.6	37.4	21.0
DoD—PACOM	35.7	24.2	13.0
DoD—EUCOM	35.2	24.7	10.9
DoD—JFCOM	30.7	22.2	9.1
DoD—U.S. Marine Corps	30.3	28.8	21.8
DoD—SOCOM	28.7	18.8	9.8
DoD—NORTHCOM	26.6	18.0	6.6
DoD—STRATCOM	25.9	18.0	8.4
DoD—Defense Intelligence Agency	25.4	23.6	7.0
Non-U.S. military	25.2	23.7	20.4

Frequent Interactions with Organizations: Differences Across Major Billet Organizations

Table 4.3 shows the percentage of officers, categorized by the major billet organization to which they were assigned, who reported frequent interactions with a subset of agencies, services, and organizations. As stated above, respondents were asked about their interactions with 73 different organizations. To keep the table to a manageable size, we selected billets in one organization that represents a high degree of jointness—the Joint Staff. We then examined the responses of officers serving in Joint Staff billets and selected all those organizations (n = 19) with which at least 25 percent of the officers interacted frequently. This subset encompasses the 15 organizations in Table 4.2, with the exception of non-U.S. military. We included those organizations in Table 4.3 for the sake of completeness.

The percentage of officers in Joint Staff billets who interact frequently with other organizations is considerably higher than what we saw across all JDAL billets. About seven in ten officers in the Joint Staff billets interact frequently with the OSD, Joint Staff, and the services, and five to seven out of ten officers interact frequently with the combatant commands. About 15 percent of those officers interact frequently with non-U.S. military. Officers serving in OSD billets were equally likely to interact with the Joint Staff, OSD, and the services but were much less likely to interact with the combatant commands.

Officers in CENTCOM JTF billets obviously were likely to interact with others at CENTCOM and with non-U.S. military, the Army, and the Department of State but were not as likely to be working with OSD, the Joint Staff, or the other combatant commands. In general, officers in the service-nominated billets were more likely to be interacting with their own service than with another service. However, officers in Marine Corps billets reported

Table 4.3
Percentage of Officers in Major Billet Organizations Reporting Frequent Interactions with Selected Organizations

	Joint Staff	OSD	CENTCOM JTF	Army	Navy	Air Force	Marine Corps	International Organizations
DoD—Joint Chiefs of Staff	97.9	90.2	14.6	8.1	11.9	6.2	40.7	8.9
DoD—Office of the Secretary of Defense	89.9	98.0	17.9	8.7	15.4	14.7	32.2	7.6
DoD—U.S. Army	82.0	77.3	41.8	67.0	24.9	24.6	52.5	15.3
DoD—U.S. Air Force	76.9	79.2	23.5	28.6	42.8	75.0	44.1	32.6
DoD—U.S. Navy	74.3	77.3	23.7	16.3	87.4	19.5	62.7	18.6
DoD—U.S. Marine Corps	69.0	65.9	24.1	16.7	51.2	11.1	89.8	4.2
DoD—CENTCOM	66.9	32.2	72.9	26.7	14.8	19.3	45.8	8.9
DoD—JFCOM	65.8	31.8	13.2	9.1	10.7	8.1	28.8	9.3
DoD—EUCOM	62.8	26.7	11.7	14.0	10.9	8.4	30.5	24.2
DoD—PACOM	61.6	28.6	1.7	11.5	21.1	9.9	42.4	0.4
DoD—NORTHCOM	58.3	23.1	1.4	5.5	9.4	6.0	18.6	3.0
DoD—SOCOM	52.6	22.0	9.3	13.1	7.5	8.4	18.6	0.4
DoD—SOUTHCOM	51.4	19.6	1.4	7.7	6.3	5.8	20.3	5.5
DoD—STRATCOM	50.4	22.8	2.9	3.8	8.0	11.7	25.4	0.0
DoD—TRANSCOM	47.4	14.5	2.9	6.3	4.6	8.6	20.3	0.9
DoD—Defense Intelligence Agency	30.2	31.8	13.1	7.2	8.9	6.0	3.4	3.4
Department of State	29.3	27.5	36.3	4.2	3.2	2.6	3.4	5.5
DoD—U.S. Army National Guard	29.2	22.0	15.3	19.0	2.8	3.7	5.1	0.4
DoD—SPACECOM	25.6	11.0	0.2	2.5	3.2	10.2	13.6	0.0
Non-U.S. military	15.0	23.9	50.3	23.0	24.5	16.7	22.0	58.9

Table 4.3—Continued

	CSAs	Other Non-OSD Defense Agencies	OSD Defense Agencies	Educational Agencies	Geographic Commands	Force Provider	Functional Commands
DoD—Joint Chiefs of Staff	35.0	27.6	40.0	29.2	44.3	43.0	51.4
DoD—Office of the Secretary of Defense	39.6	36.3	80.7	23.0	35.4	30.9	34.7
DoD—U.S. Army	41.8	46.3	65.2	31.9	43.3	45.3	30.0
DoD—U.S. Air Force	44.2	50.8	65.2	30.1	40.1	44.4	48.1
DoD—U.S. Navy	40.6	45.5	63.2	28.3	37.0	43.2	33.1
DoD—U.S. Marine Corps	25.6	29.2	44.5	20.8	24.8	36.8	19.2
DoD—CENTCOM	31.1	26.7	20.0	10.2	38.1	50.8	37.1
DoD—JFCOM	16.7	20.5	12.9	17.3	22.0	82.3	31.9
DoD—EUCOM	28.2	21.5	13.6	10.2	33.0	35.2	31.6
DoD—PACOM	25.8	28.9	13.6	9.7	35.2	33.5	37.3
DoD—NORTHCOM	15.2	29.2	16.1	11.1	20.0	27.8	37.5
DoD—SOCOM	16.5	17.9	7.7	8.0	31.7	25.9	23.3
DoD—SOUTHCOM	14.0	16.0	9.0	8.4	16.8	23.5	22.1
DoD—STRATCOM	17.8	28.8	10.3	8.0	14.7	20.0	60.9
DoD—TRANSCOM	12.6	11.4	9.7	6.6	14.9	19.1	31.6
DoD—Defense Intelligence Agency	39.4	24.2	11.6	6.2	22.8	14.0	15.9
Department of State	17.0	10.5	14.2	12.0	21.2	5.8	4.9
DoD—U.S. Army National Guard	5.6	7.7	20.7	10.2	11.1	9.7	6.5
DoD—SPACECOM	7.8	10.9	5.8	4.0	5.8	7.8	21.7
Non-U.S. military	20.9	22.2	18.1	23.0	28.7	13.4	9.8

interacting frequently with all other services and organizations such as the Joint Staff, OSD, and the combatant commands. This may reflect the greater selectivity and small sample size of the billets nominated by the Marine Corps.

Officers in billets in NATO and IADB—the two organizations included under international organizations—reported frequent interactions with the Air Force and EUCOM, and with non-U.S. military and (although not shown here) NATO. Officers in billets in other organizations—CSAs, non-OSD defense agencies, and educational agencies—tended to interact frequently with their own organizations and to a lesser degree with the services and non-U.S. military. Officers serving in OSD defense agencies tended to work with OSD, JCS, and the services, as did those in combatant commands. In addition, the combatant commands interacted with other combatant commands. Among the combatant commands, geographic commands tended to have the most interactions with non-U.S. military.

Average Number of Frequent Interactions with "Non–Own-Service" Organizations

One of the possible indicators of jointness is the overall number of organizations with which an officer serving in a billet interacts frequently. Because the definition of joint duty assignment explicitly excludes assignments in an officer's own military department, it is important to exclude *own-service* interactions from the count. While the tables shown earlier do not exclude own-service interactions, we do so now. Thus, for example, for an Army officer, frequent interactions with the Army, Army Reserve, or Army National Guard would not count toward the number of frequent interactions, whereas frequent interactions with other services and other organizations would.

Table 4.4 provides the average number of non–own-service "frequent" (monthly or more frequent) interactions reported by officers in billets grouped by JDAL billet category and major billet organization. In addition, because the distributions are very skewed (a small percentage of cases reporting very large numbers of interactions), we report the median (50th percentile) and the 10th and 90th percentiles as well as the mean.[1] Thus, across all respondents, the average number of reported frequent interactions with non–own-service organizations was 4.8. The median was lower—three—so half the respondents reported interactions with three or fewer organizations, and half reported interactions with three or more organizations. At least 10 percent of the respondents reported zero frequent interactions with non–own-service organizations, while 10 percent of the respondents reported interacting frequently with 17 or more non–own-service organizations. Thus, the middle 80 percent of the distribution reported frequent interactions with 0–17 non–own-service organizations.

It is clear that the average number of frequent interactions varies considerably across JDAL billet category and major billet organizations. Table 4.4 shows that, on average, officers in JDAL billets tended to have the highest number of frequent interactions (more than seven) with non–own-service organizations. Officers in non-JDAL billets in external

[1] In general, the pth percentile of a distribution is the value below which p% of the observations lie. In this instance, the 10th, 50th, and 90th percentiles are the values (the number of non–own-service organizations with which officers interact frequently) below which 10, 50, and 90 percent of the observations lie.

Table 4.4
Distribution of Average Number of Non–Own-Service Organizations with Which Officers Interact Frequently, by JDAL Category and Major Billet Organization

	Number of Non–Own-Service Organizations		
Billet Category	**Mean**	**Median**	**10th/90th Percentiles**
Overall	4.8	3.0	0/17
JDAL Category			
JDAL billets	7.4	6.0	1/16
Non-JDAL billets in external organizations	6.1	4.0	0/15
Service-nominated billets	2.7	1.0	0/8
Major Billet Organization			
Joint Staff	13.1	13.0	2/22
OSD	10.8	9.0	3/21
OSD defense agencies	7.7	5.0	6/20
Other non-OSD defense agencies	6.7	5.0	1/16
CSAs	6.6	5.0	1/15
Force provider	6.5	5.0	0/15
Geographic commands	6.4	5.0	1/14
Functional commands	6.4	5.0	1/15
Educational agencies	5.2	3.0	0/14
U.S. Marine Corps	5.1	3.0	0/15
CENTCOM JTF	4.1	3.0	0/10
U.S. Navy	3.8	2.0	0/10
International organizations	2.5	2.0	1/4
U.S. Air Force	2.5	1.0	0/6
U.S. Army	2.4	1.0	0/7

organizations reported an average of six interactions, while service-nominated billets interacted with about three non–own-service organizations on average. The medians were somewhat smaller, ranging from one for service-nominated billets to six for JDAL billets.

Officers serving in the Joint Staff or OSD reported interacting with between 11 and 13 non–own-service organizations frequently, and the medians were equally high—13 and 9, respectively. At the upper end, 10 percent of the officers reported interacting frequently with 21 or 22 organizations.

The defense agencies, combat support agencies, and the combatant commands reported an average of six to eight non–own-service interactions, but there was no difference among them in the medians (five organizations). The 10th and 90th percentiles ranged from zero to 17 overall, with the exception of the Joint Staff, OSD, and OSD defense agencies, for which the 90th percentile was higher (22, 21, and 20, respectively).

CENTCOM JTF, educational agency, and international organization billets interacted frequently with three to five organizations on average, but the medians were lower (two to three organizations). International organization billets also had a much smaller spread, with 80 percent of officers reporting between one and four frequent interactions.

Among the services, the Marine Corps billets reported interaction with an average of five and a median of three organizations, with 10 percent of officers reporting frequent interactions with 15 or more non–own-service organizations. The Navy billets reported an average of four interactions, with 10 percent of officers reporting ten or more frequent interactions. The averages and spread were much smaller for Army- and Air Force-nominated billets, and their median was one.

Frequent Interactions with Types of Personnel: Differences by JDAL Category and Major Billet Organization

Respondents also were asked about the types of personnel with whom they interacted and the frequency of their interaction. The list of personnel from which respondents were asked to choose included nine groups: U.S. Army personnel (officer, enlisted or civilian; active duty; National Guard; or the reserves); U.S. Navy personnel (officer, enlisted or civilian; active duty; or the reserves); U.S. Air Force personnel (officer, enlisted or civilian; active duty; National Guard; or the reserves); U.S. Marine Corps personnel (officer, enlisted or civilian; active duty, or the reserves); U.S. Coast Guard personnel (officer, enlisted or civilian; active duty or the reserves); other DoD civilians; other U.S. civilians; non-U.S. civilians; and non–U.S. military officers.

As in the section above, we focus on the prevalence of frequent interactions—monthly or more frequent—rather than just any interaction. We first present all interactions, including with own-service personnel. We then calculate the average number of frequent interactions with non–own-service personnel to get closer to DoD's definition of "jointness."

Overall, between 76 and 81 percent reported having frequent interactions with Army and Air Force personnel and DoD civilians, and about two-thirds reported interacting frequently with Navy personnel. A somewhat lower percentage—56–62 percent—reported interacting with Marine Corps personnel and with U.S. civilians, 32 percent reported interacting with non–U.S. military officers, and 16–22 percent reported interacting with Coast Guard personnel and non-U.S. civilians.

Table 4.5 reports the percentages of officers, grouped by JDAL category and by major billet organization, reporting frequent interactions with the various types of personnel. More than 90 percent of officers in JDAL billets interacted with Army, Navy, and Air Force personnel and DoD civilians on a frequent basis, and about 83 percent interacted frequently with Marine Corps personnel. About three-quarters reported working with non-DoD civilians on a regular basis.[2] The percentage of officers in JDAL billets reporting frequent interactions with non–U.S. military officers is much smaller—about 43 percent—and less than 30 percent interacted frequently with Coast Guard personnel or with non-U.S. civilians. The pattern for officers serving in non-JDAL billets in external organizations is very similar, although the percentages are somewhat lower (by three to ten percentage points). By and large, the pattern of interactions is similar across the various billet organizations, although officers serving in international organizations and in CENTCOM JTF billets reported very high levels of interaction with non–U.S. military officers (95 and 81 percent, respectively). Close to one-half of those in educational agency billets reported frequent interactions with the Coast Guard as did about 40 percent of those in Joint Staff and OSD defense agency billets.

---

[2] Given this very high percentage, we suspect that respondents may have included interactions with civilian contractors when responding to this question.

Table 4.5
Percentage of Officers Reporting Frequent Interactions with Various Types of Personnel, by JDAL Category and Major Billet Organization

	Army	Navy	Air Force	Marine Corps	Coast Guard	DoD Civilian	Non-DoD Civilian	Non-U.S. Civilian	Non–U.S. Military Officer
JDAL Category									
JDAL billets	92.5	91.0	91.7	82.7	25.3	90.8	75.7	29.5	43.2
Non-JDAL billets in external organizations	89.4	87.3	89.5	73.5	19.0	91.0	68.3	25.0	33.7
Service-nominated billets	60.0	43.9	68.7	32.2	9.5	71.0	52.5	16.4	23.9
Major Billet Organization									
U.S. Army	96.0	26.6	45.1	25.7	5.6	68.7	51.0	24.1	27.2
U.S. Navy	37.5	97.5	57.2	68.8	27.2	79.5	60.2	13.7	32.3
U.S. Air Force	39.0	33.8	94.4	20.3	4.7	68.8	50.0	11.1	17.3
U.S. Marine Corps	73.8	82.0	63.9	100.0	18.0	88.5	67.2	9.8	21.3
Joint Staff	92.8	91.2	90.5	89.6	38.6	90.6	73.6	16.8	28.2
OSD	89.4	93.3	91.8	81.6	27.8	98.4	78.0	21.6	23.5
CENTCOM JTF	97.9	93.6	94.2	92.1	12.2	79.5	72.4	66.5	81.4
International organizations	70.0	77.8	94.7	37.9	0.0	51.0	52.3	82.3	95.1
CSAs	88.2	84.7	87.4	66.8	13.3	95.3	67.3	19.9	23.6
Other Non-OSD defense agencies	85.3	83.7	88.4	65.0	19.8	94.8	69.9	18.4	29.7
OSD defense agencies	86.1	88.0	87.3	68.4	39.9	89.9	63.3	12.7	12.7
Educational agencies	87.2	84.7	86.0	76.6	48.1	86.0	72.8	34.0	56.2
Geographic commands	94.3	91.6	91.6	85.8	27.6	91.8	76.8	34.8	50.7
Force provider	93.3	92.5	92.1	89.9	23.2	90.5	77.0	13.1	27.9
Functional commands	92.6	95.5	95.8	82.7	17.1	91.5	74.3	5.4	18.3

Service-nominated billets tended to report high levels of interaction with DoD civilians, Army and Air Force personnel, and, to a lesser extent, with Navy and non-DoD civilians. Indeed, as the Table 4.5 shows, the highest levels of interaction among service billets tended to be with own-service personnel, and there was a high level of interaction with DoD civilians.

Average Number of Non–Own-Service Personnel with Whom Officers Interact Frequently

For each officer, as mentioned earlier, we also calculated the number of interactions with personnel *not from the officer's own service* and then averaged these interactions across major billet organizations and JDAL categories. As was done earlier, Table 4.6 shows the means, medians, and 10th and 90th percentiles for the number of types of non–own-service personnel with whom officers interact frequently, by JDAL billet category and major billet organization. To facilitate a comparison, we show billet categories and organizations in the same order as that in Table 4.4.

On average, respondents reported interacting frequently with 3.8 types of non–own-service personnel, with a median of four different types of personnel. The numbers in the 10th and 90th percentiles were one and seven, respectively. The findings for the service-nominated billets differ significantly from those for the other two types of billets. The means ranged from 2.3 to 3.6 among the services and from 4.4 to 5.3 among non-service organizations. Overall, using the median figures, JDAL and non-JDAL billet officers reported interactions with five types of personnel, compared with two types of personnel for officers serving in the billets nominated by the services. Officers serving in CENTCOM JTF, educational agencies, and geographic commands reported interacting frequently with six types of personnel (excluding own-service personnel), while all other non-service billets reported interacting with five types of non–own-service personnel. Among the services, the median for the Navy and Marine Corps billets was four, compared with two for the Army and Air Force billets.

Summary

Using the median number of organizations and personnel with whom officers interact frequently as an indicator of jointness, we can rank the billets in the various organizations (see Table 4.7), where 1 indicates the largest number of non–own-service organizations or personnel with whom officers interact frequently. Among the JDAL categories, JDAL billets rank first on both indicators; among the major billet organizations, the Joint Staff and OSD billets rank first and second, respectively. By comparison, CENTCOM JTF billets rank low on the indicator for frequency of interactions with non–own-service organizations but very high on the indicator for frequency of interactions with non–own-service personnel. Other non-service organizations were ranked in the middle, while the services generally ranked last or next to last with respect to the two indicators.

Table 4.6
Distribution of Average Number of Non–Own-Service Personnel with Whom Officers Interact Frequently, by JDAL Category and Major Billet Organization

Billet Category	Number of Non–Own-Service Personnel		
	Mean	Median	10th/90th Percentiles
Overall	3.8	4.0	1/7
JDAL Category			
JDAL billets	5.1	5.0	2/7
Non-JDAL billets in external organizations	4.6	5.0	2/7
Service-nominated billets	2.7	2.0	2/7
Major Billet Organization			
Joint Staff	5.0	5.0	2/7
OSD	4.9	5.0	3/7
OSD defense agencies	4.5	5.0	1/7
Other non-OSD defense agencies	4.5	5.0	1/7
CSAs	4.5	5.0	2/7
Force provider	4.7	5.0	1/7
Geographic commands	5.3	6.0	2/7
Functional commands	4.5	5.0	2/6
Educational agencies	5.3	6.0	1/8
U.S. Marine Corps	3.5	4.0	0/6
CENTCOM JTF	5.2	6.0	0/7
U.S. Navy	3.6	4.0	1/6
International organizations	4.4	5.0	1/7
U.S. Air Force	2.3	2.0	0/5
U.S. Army	2.6	2.0	0/6

Table 4.7
Rankings of Billet Categories on Median Number of Interactions with Non–Own-Service Organizations and Personnel

Billet Category	Rank Based on Median Number of Interactions with Non–Own-Service Organizations	Rank Based on Median Number of Interactions with Non–Own-Service Personnel
JDAL Category		
JDAL billets	1	1
Non-JDAL billets in external organizations	2	1
Service-nominated billets	3	3
Major Billet Organization		
Joint Staff	1	4
OSD	2	4
OSD defense agencies	3	4
Other Non-OSD defense agencies	3	4
CSAs	3	4
Force provider	3	4
Geographic commands	3	1
Functional commands	3	4
Educational agencies	9	1
U.S. Marine Corps	9	12
CENTCOM JTF	9	1
U.S. Navy	12	12
International organizations	12	4
U.S. Air Force	14	14
U.S. Army	14	14

NOTE: Organizations with the same median number received the same ranking.

CHAPTER FIVE

Typical Roles and Responsibilities of Officers Serving in Joint or Potential Joint Billets

This chapter examines the set of tasks that typically characterize various billet assignments, the percentage of time spent on those tasks during a typical workweek, the relative importance of a task to a job, the level of responsibility for carrying out a task, and whether and how these characteristics differ by billet organization category and JDAL status of the billet.

Categorization of Officers' Jobs and Primary Focus of Their Work

We first begin by categorizing billets included in the survey according to whether the jobs were primarily strategic, tactical, or operational. Overall, about 39 percent of the billets in the JOM survey were described as primarily strategic, 46 percent as primarily operational, and 15 percent as primarily tactical. Not unexpectedly, the percentage mix differs depending on the JDAL status of the billet, as shown in Figure 5.1. For example, close to 60 percent of JDAL billets were described as being strategic, compared with less than a quarter of the internal service billets being described as such. Non-JDAL billets in external organizations were similar to JDAL billets in being described as primarily strategic. The largest percentage of billets dealing with tactical matters—27 percent—were in the service-nominated billets, compared with less than 6 percent in external organizations and 3 percent of JDAL billets.

Respondents were also asked to describe the primary focus of their efforts, whether operational/supportability matters pertaining to a combatant commander's AOR or to several AORs; defense-wide issues or matters that affect one or more combatant commanders, military departments, or defense agencies; or some other matter. About 44 percent of JDAL billets and 49 percent of non-JDAL billets in external organizations were focused on defense-wide issues or multiservice/multiagency matters, compared with 27 percent of internal service billets (see Figure 5.2). About a third of the respondents in internal service billets reported that they were focused on other matters.

When we examine the major billet organizations, we find distinct differences among them in the percentage of strategic, operational, and tactical billets (see Figure 5.3). For example, more than 90 percent of OSD staff billets and close to four-fifths of Joint Staff billets were described as being primarily strategic, compared with much lower percentages (between 18 and 39 percent) of internal service billets and billets in CENTCOM JTF being described as such. More than 70 percent of the billets in the

Figure 5.1
Percentage of Billets Described as Strategic, Operational, or Tactical, by JDAL Category

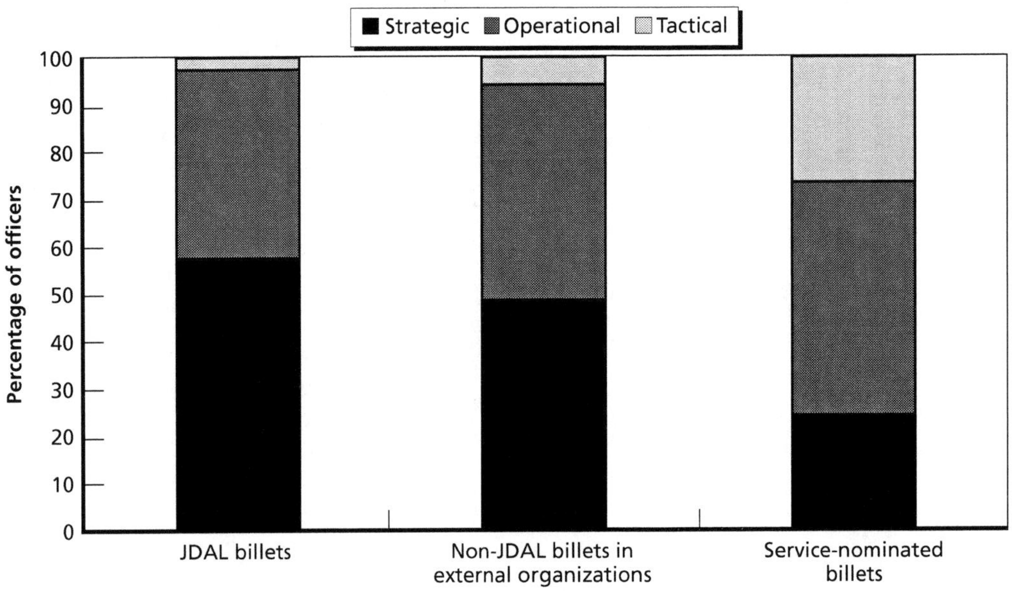

RAND *TR349-5.1*

Figure 5.2
Percentage of Billets Focused on Operational Matters, Defense-Wide/Multiservice/ Multiagency Issues, or Other Matters, by JDAL Category

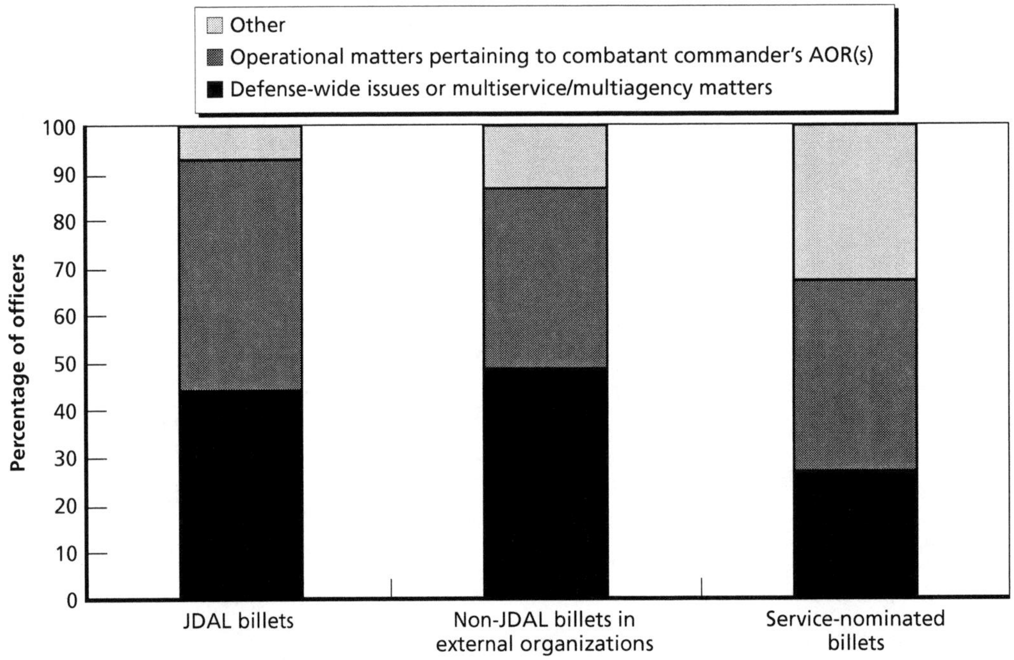

RAND *TR349-5.2*

Figure 5.3
Percentage of Billets Described as Strategic, Operational, or Tactical, by Major Billet Organization

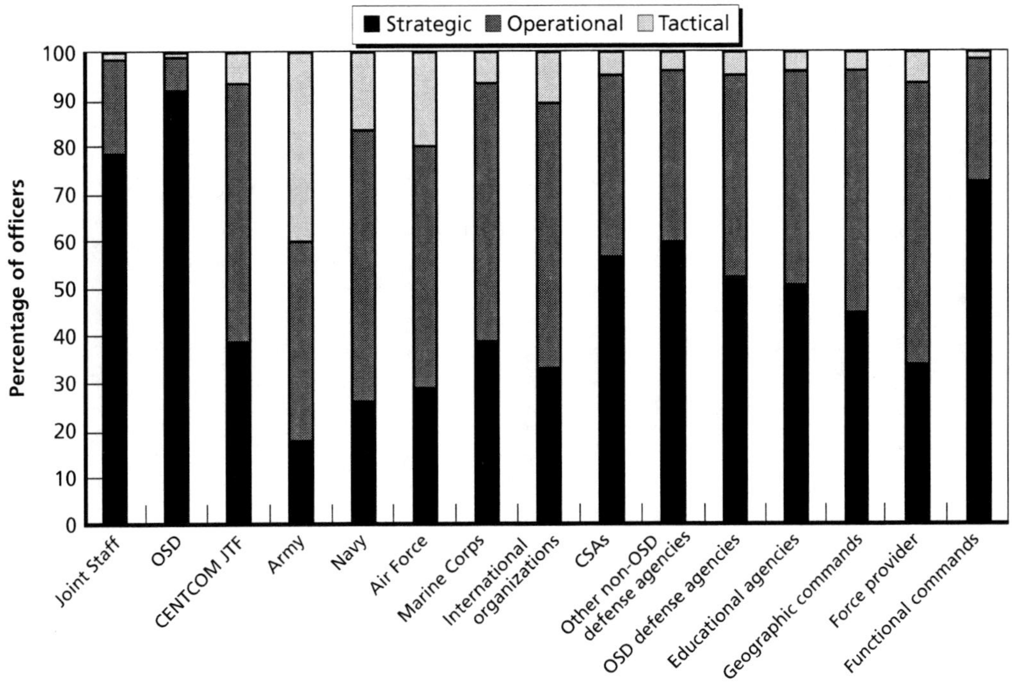

RAND *TR349-5.3*

functional commands were described as strategic, compared with 45 percent of the combatant command billets and a third of the force provider billets. About 40 percent of the internal Army billets were described as tactical—this is the highest such percentage reported across all the billet organizations and is considerably higher than that reported by the other services (17 percent of Navy billets, 20 percent of Air Force billets, and only 6 percent of Marine Corps billets). The percentage of tactical billets was very low across the remaining organizations, generally less than 10 percent.

Figure 5.4 shows the percentage of billets by major billet organization focusing on defense-wide issues or on matters cutting across agencies or departments. On this measure, OSD staff and Joint Staff billets ranked the highest followed by billets in external organizations. A little more than half of force provider and functional command billets and about half of the Marine Corps billets fell into this category. Billets in the services (with the exception of the Marine Corps), geographic commands, international organizations, and CENTCOM JTF were the least likely to be focused on defense-wide issues. More than three-quarters of the CENTCOM JTF billets and about 45 percent of billets in international organizations were focused on operational matters pertaining to combatant commanders' AORs. A substantial number of billets (between 25 and 40 percent) in some organizations (Army, Navy, Air Force, international organizations, OSD defense agencies, and educational agencies) were focused on matters other than the two discussed here.

Figure 5.4
Percentage of Billets Focusing on Defense-Wide/Multiservice/Multiagency Matters, by Major Billet Organization

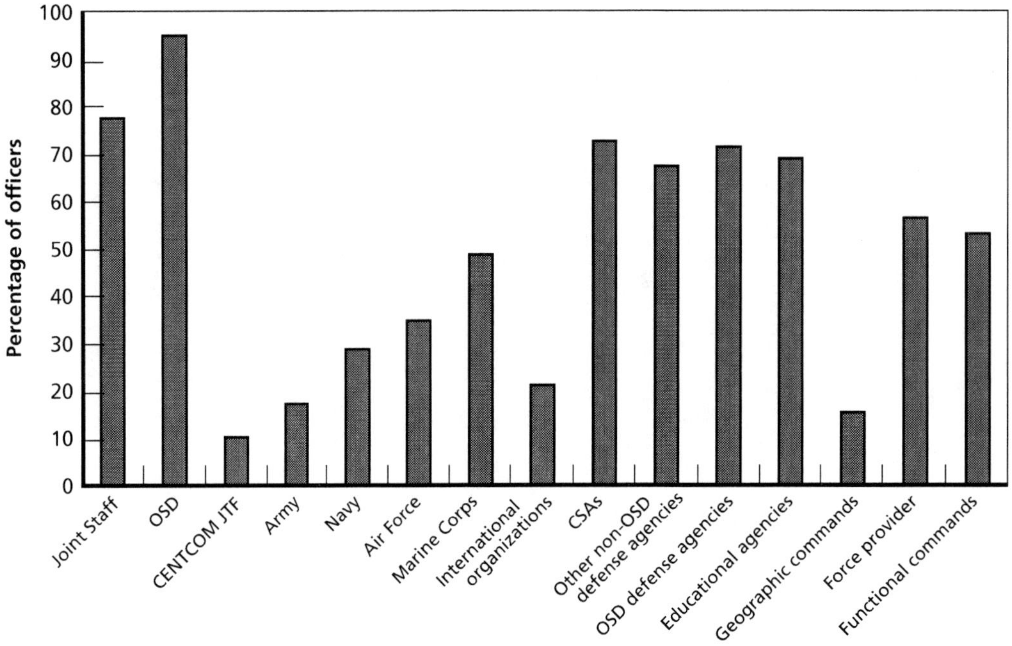

RAND *TR349-5.4*

Tasks Typically Performed in a Billet, Allocation of Time to Tasks, and Relative Importance of Tasks to Job

In this section, we examine typical tasks in a billet, allocation of time to tasks, and the relative importance of the tasks, by JDAL category and major billet organization.

JDAL Category

Tasks Typically Performed. Survey respondents were provided with a detailed list encompassing 38 separate tasks identified from various job books and from feedback from experts and were asked to indicate the tasks they typically perform. Because respondents were allowed to skip tasks they did not normally perform, we coded a skipped task as "not performed" if the respondent responded positively to at least one item in the set of 38 to distinguish those responses from the responses that were missing data for the entire set of items.

Table 5.1 shows the list of tasks that appeared in the survey and the percentage of respondents who answered affirmatively to questions about those tasks, categorized by the JDAL status of the billet. Because one of the primary goals of the follow-on work is to identify the characteristics of "jointness," we organized the table in descending order of the percentage responses from officers in JDAL billets. This setup allows us to highlight the set of tasks typically performed by those in JDAL assignments—billets that are now explicitly recognized as providing joint experience. If we set a cutoff at 20 percent, i.e., include only tasks performed by 20 percent or more of those assigned to

Table 5.1
Percentage of Respondents Who Typically Perform Certain Tasks, by JDAL Category

	JDAL Billets	Rank Order of Task	Non-JDAL Billets in External Organizations	Rank Order of Task	Service-Nominated Billets	Rank Order of Task
Provide strategic direction and integration	59.0	1	45.2	2	34.0	2
Provide administrative or technical support	50.6	2	56.4	1	49.2	1
Develop/assess joint policies	37.5	3	22.8	4	12.1	21
Develop/assess joint doctrine	32.8	4	18.8	9	12.9	20
Foster multinational, inter-agency, alliance, or regional relations	31.6	5	21.3	6	15.0	16
Provide or exercise command and control	25.8	6	19.3	8	34.0	3
Develop, conduct or provide intelligence, surveillance, and reconnaissance	21.5	7	20.8	7	17.7	14
Resource/financial management	21.2	8	22.3	5	23.0	6
Establish theater force requirements and readiness	15.0	9	17.7	11	7.6	26
Acquisition/joint program management	14.9	10	8.3	17	23.2	5
Provide logistics or combat service support	14.6	11	23.7	3	22.4	7
Research, development, testing, evaluation, and simulations	13.6	12	17.9	10	18.6	12
Provide or coordinate protection of the force, or protect the force	12.9	13	13.9	12	20.7	10
Operations other than war	12.9	13	12.4	13	18.0	13
Special operations	12.8	15	9.3	15	10.4	25
Conduct deployment, redeployment, movement, or maneuver of forces	12.7	16	6.1	22	23.3	4
Provide sustainment	12.2	17	7.0	20	21.8	8
Employ forces	8.6	18	5.9	23	19.1	11
Sustain theater forces' communications, and computers (C4)	8.0	19	5.3	25	5.5	31
Conduct force development	8.0	19	5.3	25	11.1	23
Host-nation security	7.8	21	4.8	28	4.6	32
Counter or manage deterrence of CBRNE weapons, or operate in a CBRNE environment	6.4	22	4.0	30	4.0	34
Deploy and maneuver forces	6.3	23	3.6	32	16.9	15
Safety	6.1	24	8.2	18	21.0	9
Legal affairs	5.9	25	7.9	19	13.7	19
Targeting of enemy information systems	5.7	26	4.1	29	7.0	27
Engineering	4.9	27	9.4	14	11.2	22
Coordinate counterproliferation in theater	4.8	28	5.1	27	1.4	38
Conduct mobilization	4.8	28	2.2	35	11.1	24
Civil affairs and psychological operations	4.3	30	3.7	31	6.0	29

Table 5.1—Continued

	JDAL Billets	Rank Order of Task	Non-JDAL Billets in External Organizations	Rank Order of Task	Service-Nominated Billets	Rank Order of Task
Mapping, charting, and geodesy	3.8	31	6.6	21	3.6	36
Maintenance	3.6	32	5.3	26	14.1	18
Employ firepower or other assets	3.3	33	3.6	33	14.3	17
Inspector General activities	3.2	34	1.7	37	5.9	30
Law enforcement	2.3	35	2.1	36	3.7	35
Medical/health services	1.9	36	3.5	34	6.6	28
Industrial management	1.8	37	5.7	24	4.5	33
Combat engineering	1.1	38	0.8	38	1.9	37

NOTE: Tasks with equal percentages of respondents performing those tasks received the same ranking.

JDAL billets, the set of tasks typically performed in JDAL assignments would encompass the following eight tasks (in rank order):

1. Provide strategic direction and integration
2. Provide administrative or technical support
3. Develop/assess joint policies
4. Develop/assess joint doctrine
5. Foster multinational, interagency, alliance, or regional relations
6. Provide or exercise command and control
7. Develop, conduct or provide intelligence, surveillance, and reconnaissance
8. Resource/financial management.

Between 50 and 60 percent of officers in JDAL billets reported that they provided (1) strategic direction and integration and (2) administrative or technical support. While the percentage rate for the former task makes sense for a joint billet, the rate for the latter task seems surprisingly high. It appears that respondents interpreted the questions regarding that task more broadly than was intended. About a third or more of the respondents reported that they undertook tasks one would expect in joint billets—developing/assessing joint policies or joint doctrine and fostering multinational, interagency, alliance, or regional relations. Twenty-one of the 38 tasks were typically performed by less than 10 percent of officers in JDAL assignments.

It is useful to compare the rankings we obtained using the indicator "percentage performing task" for JDAL billets, non-JDAL billets in external organizations, and service-nominated billets. There is considerable agreement in the rank order obtained for JDAL billets and billets in external organizations in terms of top-ranked tasks, although the percentage of officers undertaking development/assessment of joint policies or joint doctrine and fostering multinational, interagency, or regional relations was smaller in external organization billets than in the JDAL billets. This is clearly seen in Figure 5.5. Only 12–15 percent of officers in service-nominated billets undertake these more typical "joint" tasks. Approximately 20 percent of officers, regardless of the JDAL status of the billet, reported that they performed resource/financial management.

Figure 5.5
Percentage of Respondents Who Typically Perform Eight Top-Ranked Tasks, by JDAL Category

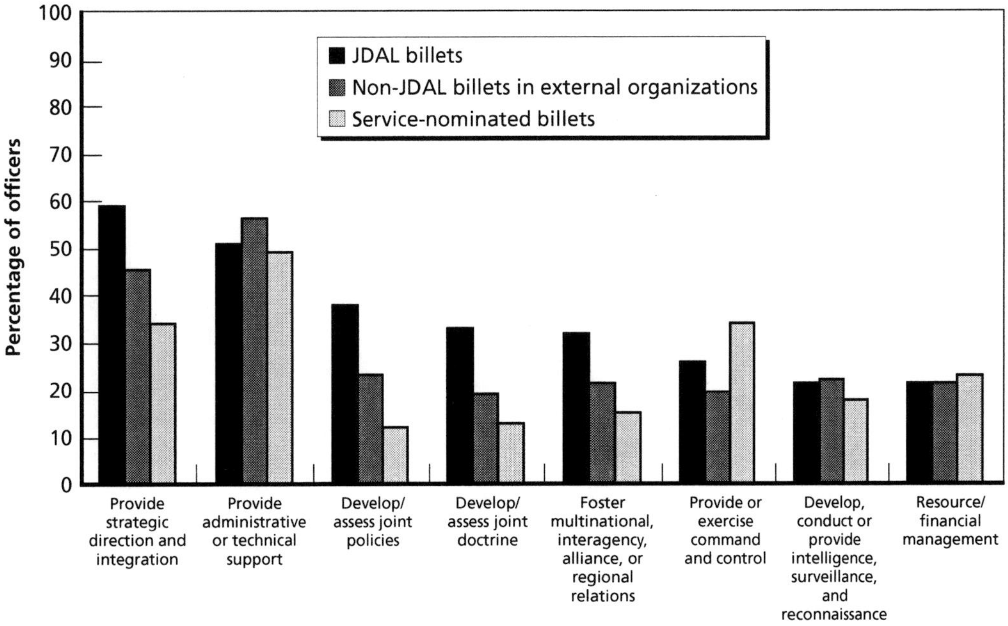

RAND *TR349-5.5*

Among service-nominated billets, in contrast to the other billets, conducting deployment, redeployment, movement, or maneuver of forces; providing or coordinating protection of the force, or protecting the force; providing sustainment; and safety ranked among the top ten tasks.

Overall, respondents reported doing an average of five tasks, with the median at four tasks. Respondents holding internal service billets had the largest mean number of tasks—5.5—compared with 5.0 for those in JDAL billets and 4.5 for those in non-JDAL billets in external organizations. In particular, those in Army and Marine Corps billets reported an average of 6.5 to 6.6 tasks during a typical workweek. Those holding Joint Staff and OSD billets performed an average of 4.2 to 4.3 tasks, somewhat lower than the average number, while those in billets in CENTCOM JTF, international agencies, and geographic commands averaged more than five tasks. Those in educational agency billets reported the lowest number of tasks on average, 3.9.

Allocation of Time to Tasks. Respondents were asked about the percentage of time they typically spend on the tasks that they perform. Because the means tended to be sensitive to outliers, we decided to use median percentages of time instead. These medians are shown in Table 5.2 for the three JDAL categories of respondents who reported performing a specific task. However, the median percentages cannot be added across tasks to represent 100-percent allocation of time during a typical workweek. All the table shows is that among officers performing a given task, half reported doing a task

Table 5.2
Median Percentage of Time Spent on Given Tasks During a Typical Workweek, by JDAL Category

	JDAL Billets	Non-JDAL Billets in External Organizations	Service-Nominated Billets
Provide strategic direction and integration	20.0	20.0	16.7
Provide administrative or technical support	20.0	22.2	20.0
Develop/assess joint policies	10.0	9.1	6.3
Develop/assess joint doctrine	9.0	8.3	6.7
Foster multinational, interagency, alliance, or regional relations	16.7	10.0	7.1
Provide or exercise command and control	10.9	12.5	16.7
Develop, conduct or provide intelligence, surveillance, and reconnaissance	31.1	33.3	11.9
Resource/financial management	10.0	10.0	8.3
Establish theater force requirements and readiness	8.3	7.1	7.1
Acquisition/joint program management	16.7	25.0	40.0
Provide logistics or combat service support	10.0	12.5	9.4
Research, development, testing, evaluation, and simulations	10.0	15.6	10.9
Provide or coordinate protection of the force, or protect the force	6.7	7.1	5.6
Operations other than war	7.1	7.7	6.5
Special operations	14.3	9.2	10.0
Conduct deployment, redeployment, movement, or maneuver of forces	8.3	7.1	7.1
Provide sustainment	8.6	9.7	8.3
Employ forces	6.7	6.3	7.7
Sustain theater forces' communications, and computers (C4)	16.7	16.4	6.4
Conduct force development	7.7	5.7	8.3
Host-nation security	6.7	5.8	3.3
Counter or manage deterrence of CBRNE weapons, or operate in a CBRNE environment	8.3	10.0	2.5
Deploy and maneuver forces	6.7	5.0	7.1
Safety	2.7	4.0	5.0
Legal affairs	4.3	6.0	4.6
Targeting of enemy information systems	7.4	7.3	6.0
Engineering	10.0	12.2	10.0
Coordinate counterproliferation in theater	5.5	4.8	2.6
Conduct mobilization	4.3	6.7	4.5
Civil affairs and psychological operations	5.2	9.1	4.8
Mapping, charting, and geodesy	4.0	10.0	4.0
Maintenance	4.4	6.0	7.4
Employ firepower or other assets	5.4	5.0	7.1
Inspector General activities	3.6	6.2	3.3
Law enforcement	5.0	5.3	3.2
Medical/health services	2.5	16.7	4.6
Industrial management	5.0	6.7	4
Combat engineering	6.0	4.2	2.9

a given percentage of time.[1] Because officers perform different tasks, the sample size for calculating the median differs across tasks.

Among respondents in JDAL billets who performed these tasks, developing, conducting, or providing intelligence, surveillance, and reconnaissance occupied close to one-third of their time on average, compared with 20 percent of their time spent on providing strategic direction and integration or providing administrative/technical support; 17 percent of their time on fostering multinational, interagency, alliance, or regional relations; 14 percent on operations other than war; and 10 percent (or less) spent on developing/assessing joint policies or joint doctrine. This suggests that billet assignments require that officers undertake a variety of tasks, and it is rare that one or two tasks predominate. However, in some positions, officers spend a substantial amount of time on one or two tasks.

Officers in non-JDAL billets in external organizations seem to allocate their time across tasks in a fashion similar to those holding JDAL billets, although they spend more of their time (25 percent) on acquisition or joint program management.

Officers in billets internal to the services spend a great deal of their time on acquisition or joint program management. About half reported spending 40 percent or more of their time on this task. Like the officers in the other billet categories, they devote 20 percent of time to providing strategic direction and integration and about 10 percent to special operations. Unlike the others, they spend far less time on intelligence activities and theater communications.

Officers Performing "Highly Joint" Tasks. Just as an example of the kinds of information that could be derived from further analysis of these data, we selected all officers doing one or more tasks that are arguably "highly joint"—(1) providing strategic direction and integration; (2) developing/assessing joint policies; (3) developing/assessing joint doctrine; and (4) fostering multinational, interagency, alliance, or regional relations. For each officer, we obtained the total percentage of time spent by the officer on these joint tasks by adding the percentages of time spent on each task across the four tasks.

Figure 5.6 shows the percentages of officers by billet category performing none, one, two, or three or more of these joint tasks. Close to 80 percent of JDAL officers are performing one or more of these tasks, and 27 percent are performing at least three of these tasks. In contrast, only 45 percent of officers in internal service billets perform any of these joint tasks, and less than 10 percent perform three or more of these tasks. The non-JDAL, non-service billets rank in the middle, with 60 percent of officers performing at least one of these tasks and 15 percent performing at least three joint tasks.

Half of all officers in JDAL billets reported doing this set of tasks 37 percent of the time during a typical workweek, compared with 27 percent of the time for officers in non-JDAL billets in external organizations and 20 percent of the time for officers in service-nominated billets. About 10 percent of those on JDAL billets reported doing these tasks 100 percent of the time.

---

[1] We tried to define a typical officer, e.g., an O-6 working in a particular organization, and tried to estimate the allocation of time to given set of tasks. Unfortunately, because officers are doing a variety of tasks, the percentage of time varies greatly depending on the number of tasks.

Figure 5.6
Percentage of Officers Performing None, One, Two, or Three or More of the Tasks Characterized as "Highly Joint," by JDAL Category

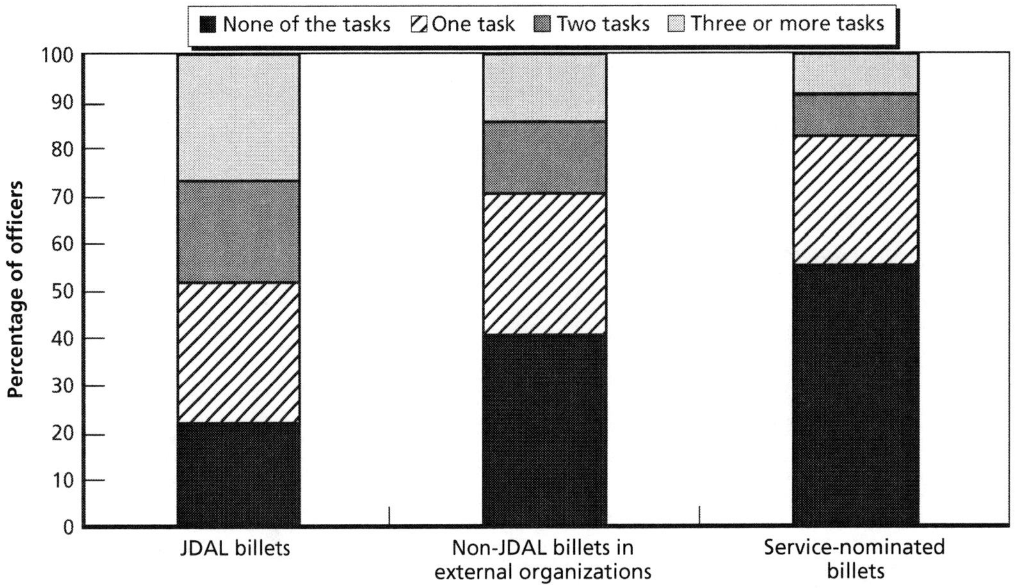

RAND *TR349-5.6*

Relative Importance of Tasks to a Job. Respondents were asked to rate the relative importance of the tasks they typically performed on a scale of 1 to 4, where 1 is peripheral importance, 2 is secondary importance, 3 is primary importance, and 4 is vitally important. We combined the latter two categories. The percentage of respondents who consider a task to be of primary/vital importance to their job is shown in Table 5.3 for the set of 38 tasks, ranked by percentage of JDAL billet respondents. The denominator in these percentages is limited to those who perform the task, so these numbers do not represent overall percentages. The table also provides the rank order for the other two groups of officers for comparison purposes.

The two top-ranked tasks for officers in JDAL assignments are providing strategic direction and integration and developing, conducting, or providing intelligence, surveillance, and reconnaissance. Three-quarters of officers who perform these activities consider them to be of primary or vital importance. These tasks were also highly ranked by officers in non-JDAL billets in external organizations and in service-nominated billets. However, there are some differences among the rankings that may be of importance when considering similarities in the work being done by officers assigned to different billets.

Among JDAL officers, between 60 and 70 percent ranked the following as important to their job: special operations; foster multinational, interagency, alliance, or regional relations; sustain theater forces' communications, and computers (C4); and provide or exercise command and control. Developing/assessing joint policies ranked lower—less than half of the officers who performed this task rated it as of primary/vital importance. Developing/assessing joint doctrine ranked even farther behind, with less than 40 percent rating it as important.

Table 5.3
Percentage of Respondents Reporting that a Task Is of Primary/Vital Importance to Their Job, by JDAL Category

	JDAL Billets	Rank Order of Tasks	Non-JDAL Billets in External Organizations	Rank Order of Tasks	Service-Nominated Billets	Rank Order of Tasks
Provide strategic direction and integration	76.1	1	72.0	2	75.0	4
Develop, conduct, or provide intelligence, surveillance, and reconnaissance	74.4	2	76.0	1	63.4	8
Special operations	67.9	3	53.1	13	66.6	7
Foster multinational, interagency, alliance, or regional relations	66.3	4	55.6	11	49.3	21
Sustain theater forces' communications, and computers (C4)	63.6	5	70.7	3	56.5	13
Provide or exercise command and control	61.3	6	63.5	5	77.9	1
Acquisition/joint program management	58.0	7	70.6	4	76.6	3
Deploy and maneuver forces	56.7	8	53.0	14	73.0	6
Employ forces	56.3	9	57.8	8	73.5	5
Conduct deployment, redeployment, movement, or maneuver of forces	54.7	10	51.4	18	60.8	10
Counter or manage deterrence of CBRNE weapons, or operate in a CBRNE environment	54.2	11	62.0	7	37.4	28
Provide logistics or combat service support	54.1	12	49.7	19	61.0	9
Establish theater force requirements and readiness	53.2	13	49.1	20	55.2	16
Employ firepower or other assets	53.0	14	48.6	21	77.5	2
Targeting of enemy information systems	52.8	15	53.3	12	52.4	19
Provide sustainment	51.7	16	47.1	23	56.1	15
Engineering	49.8	17	57.4	9	47.1	25
Research, development, testing, evaluation, and simulations	48.4	18	57.3	10	51.9	20
Develop/assess joint policies	47.7	19	45.5	25	36.2	30
Provide or coordinate protection of the force, or protect the force	47.1	20	53.0	15	56.8	12
Resource/financial management	46.2	21	48.5	22	53.8	17
Provide administrative or technical support	45.1	22	52.2	17	47.5	24
Civil affairs and psychological operations	42.3	23	52.9	16	48.3	23
Coordinate counterproliferation in theater	41.5	24	33.7	37	31.1	34
Operations other than war	41.3	25	37.3	35	44.6	26
Conduct force development	41.2	26	44.2	27	56.2	14
Combat engineering	39.3	27	39.2	31	36.9	29
Develop/assess joint doctrine	39.1	28	38.7	32	35.3	31
Host nation security	38.1	29	38.5	33	35.1	32
Conduct mobilization	37.8	30	45.3	26	43.6	27
Safety	33.5	31	42.2	29	59.0	11

Table 5.3—Continued

	JDAL Billets	Rank Order of Tasks	Non-JDAL Billets in External Organizations	Rank Order of Tasks	Service-Nominated Billets	Rank Order of Tasks
Maintenance	31.0	32	38.0	34	52.8	18
Industrial management	27.9	33	36.8	36	33.0	33
Law enforcement	27.6	34	26.4	38	28.3	36
Mapping, charting, and geodesy	23.2	35	42.2	30	29.1	35
Inspector General activities	22.8	36	44.2	28	27.9	37
Medical/health services	20.9	37	46.1	24	48.7	22
Legal affairs	20.8	38	62.8	6	26.3	38

NOTE: Tasks with equal percentages of respondents reporting that tasks are of primary/vital importance received the same ranking.

Among those in non-JDAL billets in external organizations, sustaining theater forces' communications and acquisition/joint program management ranked very high. Other tasks that ranked high on this scale include providing or exercising command and control, countering or managing deterrence of CBRNE weapons/operating in a CBRNE environment, and providing logistics or combat service support.

Officers serving in service-nominated billets were much more likely to rate their tasks as being important. Tasks such as providing or exercising command and control, employing forces, employing firepower or other assets, deploying and maneuvering forces, and acquisition/joint program management were mentioned as being important by 70 percent or more of officers performing these tasks.

Despite the apparent differences in how officers in the three categories rated the relative importance of the tasks, the correlations among the rankings were between 0.6 and 0.7.

Major Billet Organizations

We next turn to examining the roles and responsibilities of officers in the various major billet organizations.

Tasks Typically Performed. Tables 5.4a and 5.4b show the percentage of officers in the 15 major billet organizations who perform various tasks. The tasks are listed in the same rank order as the tasks in Tables 5.1 and 5.2 for ease of comparison. With some exceptions, notably among the internal service (i.e., service-nominated) billets, the rank order of tasks is fairly consistent across organizations and tracks well with the rank order obtained from JDAL billet respondents. Providing strategic direction and integration ranked high across all organizations as did providing administrative and technical support. About 40–54 percent of officers in the Joint Staff, OSD, educational agencies, force provider, and functional command billets reported developing or assessing joint policies and/or doctrine. Not surprisingly, almost half of the officers serving in international organizations reported fostering multinational, interagency, alliance, or regional relations.

Table 5.4a
Percentage of Respondents Who Typically Perform Given Tasks, by Major Billet Organization (group one of two)

	Joint Staff	OSD	CENTCOM JTF	Army	Navy	Air Force	Marine Corps	International Organizations
Provide strategic direction and integration	68.3	77.6	42.4	26.4	33.1	40.7	55.4	47.1
Provide administrative or technical support	47.0	53.7	56.1	45.7	56.3	48.6	69.2	58.2
Develop/assess joint policies	54.4	45.2	18.3	10.2	11.8	13.6	36.9	33.6
Develop/assess joint doctrine	44.1	22.8	11.2	12.7	11.5	13.1	41.5	25.4
Foster multinational, interagency, alliance, or regional relations	21.6	24.7	33.5	17.1	16.4	12.4	20.0	48.8
Provide or exercise command and control	14.1	3.9	23.1	51.9	18.7	25.7	24.6	37.7
Develop, conduct, or provide intelligence, surveillance, and reconnaissance	5.5	6.2	18.5	20.7	18.8	14.7	9.2	18.4
Resource/financial management	19.3	29.3	19.5	17.6	25.2	26.3	41.5	20.5
Establish theater force requirements and readiness	14.5	7.7	15.6	7.2	7.8	7.6	18.5	15.6
Acquisition/joint program management	15.8	30.5	8.9	6.2	23.4	37.8	12.3	7.0
Provide logistics or combat service support	8.5	5.8	25.4	30.9	15.0	18.4	32.3	16.8
Research, development, testing, evaluation, and simulations	9.0	18.2	4.5	9.1	23.7	24.7	15.4	13.5
Provide or coordinate protection of the force or protect the force	6.6	3.9	19.8	37.4	11.7	10.6	15.4	11.9
Operations other than war	5.0	3.1	9.1	17.2	5.6	6.9	3.1	2.5
Special operations	9.6	4.6	24.1	23.9	23.5	10.7	9.2	20.1
Conduct deployment, redeployment, movement, or maneuver of forces	9.1	2.3	12.8	38.4	11.1	15.8	24.6	15.6
Provide sustainment	7.0	5.0	21.1	27.7	13.7	20.1	38.5	13.1

Table 5.4a—Continued

	Joint Staff	OSD	CENTCOM JTF	Army	Navy	Air Force	Marine Corps	International organizations
Employ forces	4.8	0.4	11.8	32.5	10.8	11.5	3.1	15.6
Sustain theater forces' communications, and computers (C4)	7.0	2.7	9.1	5.5	5.1	5.5	10.8	5.3
Conduct force development	8.9	8.9	11.3	11.3	10.5	10.9	24.6	10.3
Host-nation security	2.5	3.1	16.1	8.1	3.0	2.4	1.5	7.4
Counter or manage deterrence of CBRNE weapons, or operate in a CBRNE environment	4.0	3.9	1.1	5.2	2.9	3.5	3.1	2.5
Deploy and maneuver forces	4.0	1.2	6.7	29.5	8.7	9.8	9.2	8.2
Safety	4.4	3.5	11.6	29.2	24.4	12.4	18.5	7.8
Legal affairs	7.6	13.9	5.7	17.9	15.7	9.0	15.4	6.6
Targeting of enemy information systems	2.6	0.8	6.2	10.2	8.3	3.8	3.1	3.7
Engineering	3.1	6.6	8.6	3.8	14.2	16.2	13.9	5.3
Coordinate counterproliferation in theater	2.3	1.2	2.6	1.8	1.2	1.1	1.5	0.8
Conduct mobilization	3.3	1.9	4.9	13.4	6.9	11.0	12.3	9.0
Civil affairs and psychological operations	2.9	1.2	11.6	13.5	2.3	1.1	7.7	3.7
Mapping, charting, and geodesy	1.4	0.4	5.9	4.2	3.3	3.2	4.6	4.9
Maintenance	2.5	1.2	7.5	23.1	11.8	7.1	23.1	4.5
Employ firepower or other assets	1.6	7.0	1.6	5.4	4.7	6.9	12.3	2.1
Inspector General activities	0.9	0.4	4.2	28.8	5.0	6.2	0.0	9.8
Law enforcement	1.6	2.3	5.3	5.1	2.1	3.3	4.6	1.2
Medical/health services	2.6	7.7	2.6	8.0	4.2	6.3	12.3	1.6
Industrial management	1.0	5.0	2.1	3.2	7.0	4.4	7.7	1.2
Combat engineering	0.8	0.4	2.7	2.4	1.1	1.7	9.2	2.1

Table 5.4b
Percentage of Respondents Who Typically Perform Given Tasks, by Major Billet Organization (group two of two)

	CSAs	Other Non-OSD Defense Agencies	OSD Defense Agencies	Educational Agencies	Geographic Commands	Force Provider	Functional Commands
Provide strategic direction and integration	46.3	54.7	57.1	49.8	52.5	50.7	62.3
Provide administrative or technical support	54.1	58.9	60.9	53.6	49.5	54.2	57.7
Develop/assess joint policies	18.8	25.4	31.7	40.5	32.1	40.1	41.0
Develop/assess joint doctrine	14.8	21.4	20.5	43.5	28.6	46.0	38.6
Foster multinational, interagency, alliance, or regional relations	25.1	27.5	9.3	24.1	34.5	19.6	8.7
Provide or exercise command and control	19.3	17.0	21.7	12.2	27.9	27.4	29.4
Develop, conduct, or provide intelligence, surveillance, and reconnaissance	32.1	29.2	9.3	4.6	24.0	13.5	13.4
Resource/financial management	20.5	24.4	23.6	24.5	22.4	16.4	17.5
Establish theater force requirements and readiness	4.2	8.5	2.5	3.0	19.2	11.6	8.1
Acquisition/joint program management	29.7	31.6	22.4	18.6	11.9	11.6	15.4
Provide logistics or combat service support	21.7	9.1	7.5	9.7	17.5	10.6	13.7
Research, development, testing, evaluation, and simulations	19.2	30.4	18.6	25.7	9.7	29.0	15.7
Provide or coordinate protection of the force, or protect the force	8.2	9.4	7.5	4.2	16.3	6.1	7.5
Operations other than war	3.1	3.3	2.5	3.0	21.6	7.1	2.2
Special operations	9.0	7.4	5.6	8.4	18.2	8.6	5.4
Conduct deployment, redeployment, movement, or maneuver of forces	5.0	2.4	3.1	2.1	15.6	12.9	9.1
Provide sustainment	15.3	8.5	9.3	6.8	15.3	9.6	12.6

Table 5.4b—Continued

	CSA	Other Non-OSD Defense Agencies	OSD Defense Agencies	Educational Agencies	Geographic Commands	Force Provider	Functional Commands
Employ forces	3.3	2.7	7.5	3.0	11.7	5.9	6.8
Sustain theater forces' communications, and computers (C4)	7.1	6.0	3.1	4.2	10.5	8.4	8.9
Conduct force development	2.0	5.0	3.7	5.9	8.1	14.5	5.5
Host-nation security	5.2	4.1	1.9	0.8	10.3	1.6	0.8
Counter or manage deterrence of CBRNE weapons, or operate in a CBRNE environment	7.9	2.7	4.4	1.7	6.1	2.0	8.1
Deploy and maneuver forces	1.9	0.7	3.7	3.4	8.7	5.1	4.8
Safety	9.2	4.3	11.2	2.5	7.3	3.9	5.0
Legal affairs	5.2	8.8	9.9	5.1	7.8	3.5	3.9
Targeting of enemy information systems	3.0	6.4	0.6	3.4	6.8	3.7	6.5
Engineering	10.2	13.2	4.4	3.0	5.1	4.7	4.1
Coordinate counterproliferation in theater	4.7	1.4	0.0	1.3	5.9	1.2	1.2
Conduct mobilization	3.1	1.8	5.0	0.8	7.1	8.2	4.0
Civil affairs and psychological operations	1.3	0.5	1.2	5.1	6.8	3.7	1.0
Mapping, charting, and geodesy	3.4	17.3	1.2	0.8	4.9	3.5	3.8
Maintenance	5.6	1.8	6.8	2.5	5.2	2.4	1.9
Employ firepower or other assets	2.9	2.8	24.8	0.8	3.9	5.1	1.4
Inspector General activities	0.3	0.7	0.0	1.7	4.2	2.0	3.9
Law enforcement	0.9	1.0	5.0	0.8	3.5	1.2	1.1
Medical/health services	2.5	1.1	29.2	2.1	4.0	2.7	2.8
Industrial management	5.3	2.7	3.7	2.5	1.8	0.6	1.2
Combat engineering	0.2	0.1	0.0	0.8	1.5	1.6	0.2

As we had done earlier, we selected the eight top-ranked tasks for JDAL billets and examined the percentage of officers in selected billet organizations and services who performed those tasks. The results of this analysis are shown in Figure 5.7, which as a benchmark also shows the percentage of JDAL officers performing these tasks.

The percentages of Joint Staff and OSD staff officers performing these tasks look relatively similar. They were much more likely than others to report providing strategic direction and integration and developing/assessing joint policies and were far less likely to report providing command and control or undertaking intelligence activities. The results for the CENTCOM JTF billets look more like those for the internal service billets than those for the JDAL billets.

Allocation of Time to Tasks. Tables 5.5a and 5.5b show the median percentage of time spent on certain tasks that was reported by officers in the various billet organizations who perform these tasks. Here are some highlights:

- Providing strategic direction and integration: Officers in almost every organization reported spending at least one-fifth of their time on this task, particularly officers in billets in the Joint Staff (30 percent), educational agencies and OSD staff (27–28 percent), and functional commands (25 percent).
- Sustaining theater forces' communications: Officers in Joint Staff, CSA, CENTCOM JTF, and Marine Corps billets spend between 21–24 percent time on this task.

Figure 5.7
Percentage of Respondents Who Perform Selected Tasks, by JDAL Billets, Billets in Selected External Organizations, and Service-Nominated Billets

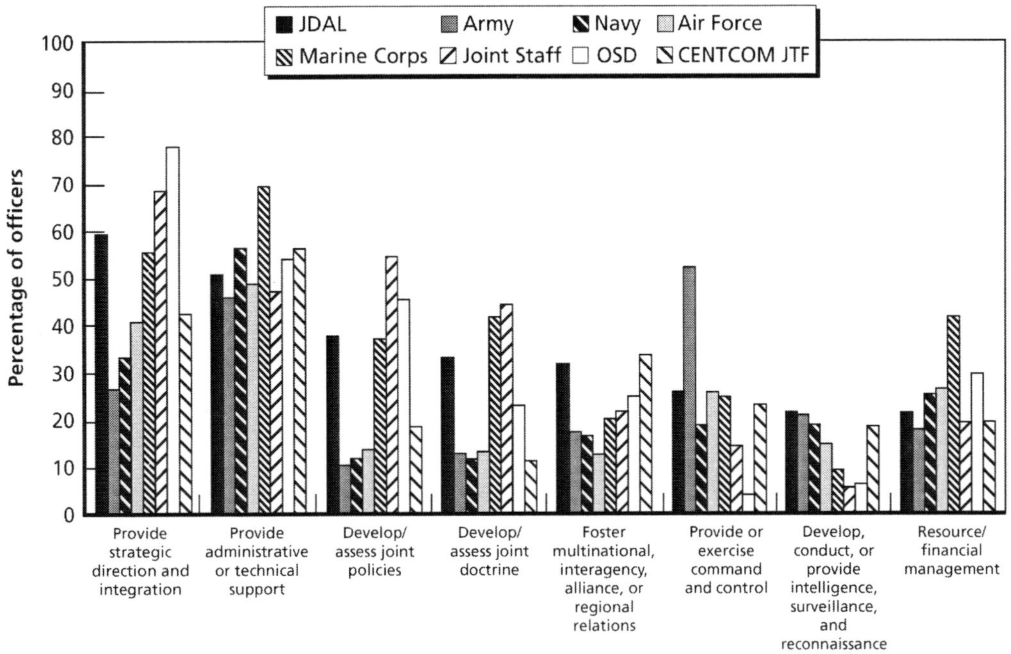

RAND *TR349-5.7*

Table 5.5a
Median Percentage of Time Spent on Given Tasks During a Typical Workweek, by Major Billet Organization (group one of two)

	Joint Staff	OSD	CENTCOM JTF	Army	Navy	Air Force	Marine Corps	International Organizations
Provide strategic direction and integration	29.5	27.3	18.8	14.3	16.7	17.5	21.4	20.0
Provide administrative or technical support	23.8	20.0	23.7	16.7	22.2	20.0	16.7	25.0
Develop/assess joint policies	16.0	15.4	6.7	6.3	8.3	6.0	11.0	9.1
Develop/assess joint doctrine	10.0	9.1	5.6	6.3	8.0	6.7	7.5	7.5
Foster multinational, interagency, alliance, or regional relations	16.1	14.3	12.5	6.7	8.3	7.7	4.7	20.0
Provide or exercise command and control	10.2	9.5	12.5	20.0	11.1	14.3	4.8	18.0
Develop, conduct, or provide intelligence, surveillance, and reconnaissance	13.3	5.0	25.0	8.3	18.2	16.7	10.2	20.0
Resource/financial management	10.6	15.4	8.0	5.6	10.0	10.0	8.3	10.0
Establish theater force requirements and readiness	8.3	7.8	8.7	6.5	8.3	8.0	15.5	8.7
Acquisition/joint program management	16.7	27.3	9.8	8.3	33.3	45.5	4.3	16.0
Provide logistics or combat service support	14.3	10.6	12.5	9.7	8.3	10.0	8.3	10.5
Research, development, testing, evaluation, and simulations	10.0	11.1	5.0	8.3	11.1	11.1	7.4	14.0
Provide or coordinate protection of the force or protect the force	9.1	12.7	7.5	6.0	7.1	4.2	6.1	5.0
Operations other than war	16.0	8.9	8.9	11.0	6.1	15.0	3.3	8.1
Special operations	6.7	6.7	8.3	5.0	27.3	5.0	4.1	8.3
Conduct deployment, redeployment, movement, or maneuver of forces	9.1	6.2	8.3	6.7	7.7	8.3	7.7	9.0
Provide sustainment	8.3	9.1	10.1	7.7	7.7	9.1	8.9	9.4

Table 5.5a—Continued

	Joint Staff	OSD	CENTCOM JTF	Army	Navy	Air Force	Marine Corps	International Organizations
Employ forces	6.7	0.0	7.3	7.8	8.1	7.2	3.2	9.4
Sustain theater forces' communications, and computers (C4)	23.8	1.8	20.8	5.5	6.7	6.5	21.3	12.5
Conduct force development	10.0	8.0	7.1	6.3	10.0	8.5	8.3	8.3
Host-nation security	7.2	6.4	6.7	3.3	2.6	3.1	0.0	5.0
Counter or manage deterrence of CBRNE weapons, or operate in a CBRNE environment	8.0	16.0	1.7	2.3	2.5	3.1	15.4	3.0
Deploy and maneuver forces	12.7	5.3	5.0	7.1	7.5	6.7	4.0	6.6
Safety	3.3	4.1	3.3	4.3	9.1	4.2	3.7	2.2
Legal affairs	7.7	6.6	4.0	4.7	5.7	4.0	4.6	4.0
Targeting of enemy information systems	8.3	3.5	6.3	4.8	10.0	4.4	10.8	5.0
Engineering	6.0	7.1	14.4	3.9	10.0	10.0	5.0	20.0
Coordinate counterproliferation in theater	4.7	7.6	5.0	2.0	6.3	3.4	0.0	4.7
Conduct mobilization	7.5	6.0	4.6	4.3	5.5	5.0	4.0	4.5
Civil affairs and psychological operations	5.0	16.7	12.0	5.6	3.3	2.7	9.1	12.5
Mapping, charting, and geodesy	8.3	8.3	5.0	4.0	2.9	4.4	10.0	6.3
Maintenance	3.8	4.3	5.2	7.7	7.7	6.8	4.0	10.0
Employ firepower or other assets	4.0	4.3	4.6	2.2	2.9	4.4	5.3	5.0
Inspector General activities	6.7	1.8	6.5	7.1	6.7	6.3	0.0	5.1
Law enforcement	3.4	6.5	6.7	2.9	3.6	3.5	2.7	2.5
Medical/health services	4.0	8.3	2.3	3.3	5.5	7.7	3.1	2.9
Industrial management	3.4	4.3	2.5	3.1	6.7	3.9	2.9	2.5
Combat engineering	2.6	3.2	4.5	2.0	3.2	4.7	4.5	6.0

NOTES: The median percentages will not add to 100 percent.
The medians are based on different groups of officers within an organization who reported that they performed a given task.

Table 5.5b
Median Percentage of Time Spent on Given Tasks During a Typical Workweek, by Major Billet Organization (group two of two)

	CSAs	Other Non-OSD Defense Agencies	OSD Defense Agencies	Educational Agencies	Geographic Commands	Force Provider	Functional Commands
Provide strategic direction and integration	18.2	20.0	19.5	27.6	16.7	20.0	25.0
Provide administrative or technical support	20.0	22.2	17.6	33.3	20.0	25.0	22.2
Develop/assess joint policies	8.3	10.0	8.3	17.1	8.3	12.5	10.0
Develop/assess joint doctrine	7.5	9.1	5.5	14.0	7.3	12.0	10.0
Foster multinational, interagency, alliance, or regional relations	16.7	11.0	9.1	16.7	14.0	8.3	7.3
Provide or exercise command and control	12.9	10.0	16.7	8.9	10.0	16.5	16.7
Develop, conduct, or provide intelligence, surveillance, and reconnaissance	44.4	20.0	16.7	5.0	37.5	20.0	25.0
Resource/financial management	9.1	10.0	13.0	16.7	9.1	10.0	10.0
Establish theater force requirements and readiness	7.1	8.9	6.7	4.0	8.3	9.1	9.1
Acquisition/joint program management	25.0	29.3	16.7	47.2	12.5	10.4	22.2
Provide logistics or combat service support	15.4	9.1	8.8	10.4	10.0	8.3	10.7
Research, development, testing, evaluation, and simulations	19.2	16.7	9.1	20.0	7.5	6.7	13.7
Provide or coordinate protection of the force, or protect the force	6.7	8.3	3.7	5.1	6.7	6.1	8.3
Operations other than war	7.1	9.1	6.7	4.4	15.4	8.8	7.1
Special operations	8.3	8.0	3.6	10.0	6.6	6.7	8.3
Conduct deployment, redeployment, movement, or maneuver of forces	6.7	6.0	3.4	6.0	7.1	10.0	9.8
Provide sustainment	10.0	8.6	8.3	5.0	8.0	8.3	10.0

Table 5.5b—Continued

	CSAs	Other Non-OSD Defense Agencies	OSD Defense Agencies	Educational Agencies	Geographic Commands	Force Provider	Functional Commands
Employ forces	6.3	6.9	5.0	6.9	6.0	9.5	8.3
Sustain theater forces' communications, and computers (C4)	23.8	8.3	10.0	10.0	16.7	8.3	16.3
Conduct force development	6.0	5.5	3.8	8.7	6.0	8.9	9.6
Host-nation security	4.8	6.1	3.4	1.9	7.1	4.5	3.3
Counter or manage deterrence of CBRNE weapons, or operate in a CBRNE environment	17.4	11.4	11.4	7.5	4.2	4.6	12.0
Deploy and maneuver forces	5.0	5.6	11.9	7.4	5.6	8.2	5.0
Safety	4.0	3.5	4.0	2.0	3.1	2.3	2.9
Legal affairs	2.9	8.3	4.5	5.8	5.6	4.5	8.3
Targeting of enemy information systems	6.1	8.6	2.0	7.0	7.1	8.0	12.5
Engineering	12.5	11.1	12.5	5.7	10.0	7.2	12.0
Coordinate counterproliferation in theater	6.9	4.1	0.0	2.0	4.3	4.8	3.6
Conduct mobilization	4.0	3.5	4.0	2.2	5.0	5.4	6.2
Civil affairs and psychological operations	4.0	5.9	2.7	4.5	6.7	4.0	9.1
Mapping, charting, and geodesy	5.0	13.8	13.3	1.7	3.9	5.0	8.7
Maintenance	5.8	4.2	4.3	12.7	5.0	6.3	5.5
Employ firepower or other assets	4.4	4.4	60.0	6.6	3.3	3.3	16.7
Inspector General activities	4.0	5.0	0.0	7.1	4.1	10.3	10.0
Law enforcement	2.5	4.8	5.0	1.9	6.2	7.3	3.6
Medical/health services	10.4	33.8	20.0	9.1	4.5	22.0	33.3
Industrial management	8.3	6.9	3.1	8.3	3.6	1.5	4.6
Combat engineering	4.0	4.8	0.0	3.8	6.7	7.5	7.7

NOTES: The median percentages will not add to 100 percent.
The medians are based on different groups of officers within an organization who reported that they performed a given task.

- Acquisition or Joint Program Management: Air Force and Navy officers spend a large percentage of time on these tasks (46 and 33 percent, respectively) as do those in OSD staff billets (27 percent).
- Special Operations: Navy officers spend more than one-quarter of their time on activities related to this task.
- Developing, conducting, or providing intelligence, surveillance, and reconnaissance: Officers in billets in CENTCOM JTF and international organizations spend about one-quarter of their time on these tasks.
- Medical/health services: Officers assigned to OSD defense agency billets spend between 20–22 percent of their time on average on these tasks; the respondents largely were health professionals in the TRICARE Management Activity. Although we see officers in non-OSD defense agencies and functional commands also reporting large percentages of time spent on medical/health services, these results are driven largely by very small sample sizes of a few health professionals.
- Fostering multinational, interagency, alliance, or regional relations: Officers in international agencies spend about 20 percent of their time on activities related to this task.
- Providing administrative or technical support: Most officers reported spending about one-fifth of their time in providing these kinds of support.

Officers Performing "Highly Joint" Tasks. As we had done earlier, we examined the number of "highly joint"[2] tasks and the time spent on these tasks by officers in the various billet organizations. Eighty-five percent of officers in Joint Staff or OSD staff billets and 75 percent of officers in Marine Corps, international organization, force provider, and functional command billets performed at least one of these highly joint tasks (see Figure 5.8). Between 22 and 36 percent of officers at these organizations performed three or more of these tasks.

The median percentage of time spent on these tasks by officers in these billets was also much higher than that in other organizations—50–60 percent for the Joint Staff, OSD staff, and educational agency billets; 40 percent for the force provider and functional command billets; and 30 percent for the Marine Corps billets. Ten percent of officers with Joint Staff, OSD staff, educational agency, force provider, or functional command billets reported doing these tasks 100 percent of their time.

In contrast, only 40–50 percent of officers in Army, Navy, and Air Force billets performed at least one of these tasks, and less than 10 percent performed three or more. The median percentage of time spent on these tasks during a typical workweek ranged from 17 to 20 percent.

---

[2] As stated above, these tasks include: (1) providing strategic direction and integration; (2) developing/assessing joint policies; (3) developing/assessing joint doctrine; and (4) fostering multinational, interagency, alliance, or regional relations.

Figure 5.8
Percentage of Officers Performing None, One, Two, or Three or More "Highly Joint" Tasks, by Major Billet Organization

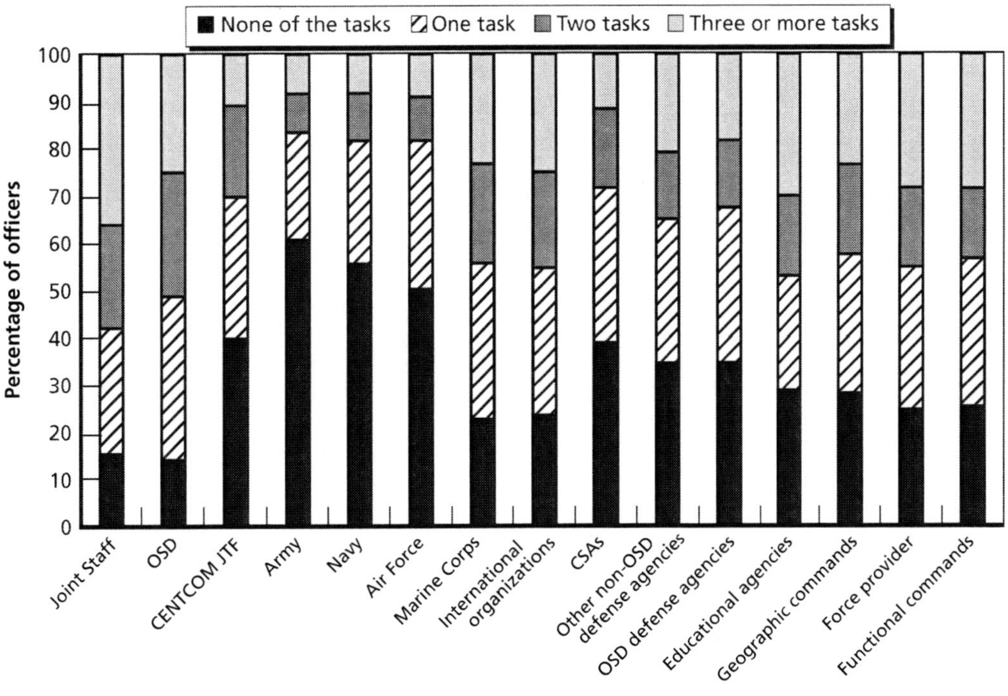

RAND *TR349-5.8*

Relative Importance of Tasks to a Job. Tables 5.6a and 5.6b show the percentage of officers performing various tasks who rated those tasks as being of primary/vital importance to their assignment. Here, we focus on the eight top-ranked tasks discussed earlier to highlight some of the similarities and differences among the responses. Figure 5.9 shows the percentages of respondents in various billet organizations who ranked these tasks as being of primary/vital importance to job performance. For purposes of comparison, the percentages for officers in JDAL billets are also illustrated for each task. Some highlights include the following:

- It is clear that officers, regardless of where they are assigned, consider providing strategic direction and integration as being important to their job. On average across organizations, 70 to 80 percent rated this task as important.
- Developing/assessing joint policies was rated highly by those in Joint Staff, OSD staff, Marine Corps, OSD defense agency, and educational agency billets.
- With the exception of those in educational agency and force provider billets, developing/assessing joint doctrine was ranked as important by substantially less than half of the respondents in each organization.
- Fostering multinational or interagency relations was rated as important by Joint Staff, OSD staff, CENTCOM JTF, international organizations, CSAs, and geographic command billets.

Table 5.6a
Percentage of Respondents Performing Certain Tasks Who Reported that the Tasks Are of Primary/Vital Importance to Their Job, by Major Billet Organization (group one of two)

	Joint Staff	OSD	CENTCOM JTF	Army	Navy	Air Force	Marine Corps	International Organizations
Provide strategic direction and integration	81.7	83.3	76.2	73.0	74.3	76.1	85.3	66.3
Provide administrative or technical support	46.0	44.5	56.0	50.7	49.0	44.4	40.4	54.6
Develop/assess joint policies	58.2	62.7	42.7	36.5	33.9	36.2	59.1	42.6
Develop/assess joint doctrine	45.3	36.9	29.4	33.0	33.8	37.7	36.0	36.0
Foster multinational, interagency, alliance, or regional relations	63.0	60.3	62.5	51.2	45.8	49.6	27.3	74.3
Provide or exercise command and control	53.3	30.0	61.6	82.4	69.2	73.1	61.6	72.9
Develop, conduct, or provide intelligence, surveillance, and reconnaissance	54.5	26.7	66.1	58.6	68.7	66.0	60.0	68.2
Resource/financial management	52.0	58.7	44.0	48.7	55.1	56.1	56.0	42.9
Establish theater force requirements and readiness	47.8	52.6	49.0	50.4	49.0	61.4	72.7	48.6
Acquisition/joint program management	54.1	67.6	56.6	52.0	74.6	80.8	33.4	58.8
Provide logistics or combat service support	75.8	35.7	60.4	64.1	57.6	57.2	88.3	63.4
Research, development, testing, evaluation, and simulations	43.7	47.9	30.7	45.1	49.6	55.5	11.1	54.9
Provide or coordinate protection of the force or protect the force	57.7	77.7	57.1	56.1	61.7	56.9	33.3	29.6
Operations other than war	70.0	71.5	43.7	65.9	54.6	72.9	0.0	33.4
Special operations	32.4	41.7	46.2	37.1	64.1	38.7	50.0	37.5
Conduct deployment, redeployment, movement, or maneuver of forces	59.7	60.0	48.7	57.7	67.1	65.8	28.5	52.7
Provide sustainment	55.5	23.1	51.6	58.7	54.6	53.3	68.2	61.3

Table 5.6a—Continued

	Joint Staff	OSD	CENTCOM JTF	Army	Navy	Air Force	Marine Corps	International Organizations
Employ forces	52.8	0.0	50.6	75.8	73.3	68.8	0.0	62.1
Sustain theater forces' communications, and computers (C4)	62.7	16.7	69.6	58.8	56.0	54.0	83.4	53.9
Conduct force development	46.5	21.7	42.9	46.1	63.4	61.6	60.0	44.0
Host-nation security	42.1	50.0	36.8	33.2	39.2	37.8	0.0	38.9
Counter or manage deterrence of CBRNE weapons, or operate in a CBRNE environment	65.7	70.0	16.7	36.3	25.9	42.9	100.0	0.0
Deploy and maneuver forces	64.5	33.3	52.4	75.1	71.5	68.7	40.0	52.7
Safety	32.3	11.1	32.4	56.7	67.2	55.7	66.7	44.5
Legal affairs	33.3	22.2	35.3	20.6	21.4	39.2	44.4	31.3
Targeting of enemy information systems	50.0	0.0	41.0	52.7	57.1	47.2	0.0	66.7
Engineering	25.0	23.5	55.8	39.2	52.5	46.6	25.0	61.6
Coordinate counterproliferation in theater	38.9	0.0	13.3	22.4	45.0	36.3	0.0	0.0
Conduct mobilization	48.0	60.0	22.6	39.9	46.3	47.0	14.3	27.3
Civil affairs and psychological operations	47.8	100.0	51.4	50.5	43.2	34.9	0.0	22.2
Mapping, charting, and geodesy	9.1	100.0	16.2	31.8	18.0	31.3	33.3	36.4
Maintenance	15.8	0.0	31.8	55.8	46.0	50.7	23.1	72.7
Employ firepower or other assets	30.8	33.3	62.5	22.3	14.3	35.0	57.1	60.0
Inspector General activities	28.6	0.0	42.3	78.4	78.1	74.1	0.0	54.6
Law enforcement	30.8	0.0	32.3	24.2	20.5	36.0	0.0	33.3
Medical/health services	40.0	45.0	28.6	43.5	44.5	56.4	28.6	25.0
Industrial management	25.0	15.4	15.4	28.0	38.3	30.6	80.0	33.3
Combat engineering	33.3	0.0	33.3	25.3	38.1	50.0	50.0	20.0

Table 5.6b
Percentage of Respondents Performing Certain Tasks Who Reported that the Tasks Are of Primary/Vital Importance to Their Job, by Major Billet Organization (group two of two)

	CSAs	Other Non-OSD Defense Agencies	OSD Defense Agencies	Educational Agencies	Geographic Commands	Force Provider	Functional Commands
Provide strategic direction and integration	68.1	74.5	77.1	81.9	73.2	75.8	79.4
Provide administrative or technical support	46.6	50.7	53.1	50.0	46.4	47.5	49.7
Develop/assess joint policies	40.0	48.9	56.8	64.3	41.1	54.9	45.7
Develop/assess joint doctrine	33.7	41.6	42.4	66.4	34.1	50.0	39.2
Foster multinational, interagency, alliance, or regional relations	65.8	53.8	40.0	50.8	65.9	44.6	41.0
Provide or exercise command and control	67.4	56.7	62.8	60.7	60.5	64.7	62.8
Develop, conduct, or provide intelligence, surveillance, and reconnaissance	83.8	67.6	73.3	40.0	76.2	65.2	66.9
Resource/financial management	42.7	50.0	55.3	50.0	46.6	51.3	46.9
Establish theater force requirements and readiness	40.7	57.8	75.0	66.7	54.2	53.6	50.0
Acquisition/joint program management	69.6	78.6	44.5	88.6	53.7	38.6	68.2
Provide logistics or combat service support	63.1	39.8	33.4	36.4	54.0	42.3	57.7
Research, development, testing, evaluation, and simulations	60.7	61.1	26.6	57.4	40.4	60.0	48.5
Provide or coordinate protection of the force, or protect the force	42.9	50.1	33.3	44.4	50.0	45.2	45.5
Operations other than war	46.3	48.3	75.0	28.6	68.8	54.6	50.0
Special operations	35.2	36.1	50.0	47.4	41.1	26.8	42.8
Conduct deployment, redeployment, movement, or maneuver of forces	50.0	35.0	60.0	75.0	51.6	63.5	56.1
Provide sustainment	57.6	56.9	33.3	14.3	50.3	44.6	62.3

Table 5.6b—Continued

	CSAs	Other Non-OSD Defense Agencies	OSD Defense Agencies	Educational Agencies	Geographic Commands	Force Provider	Functional Commands
Employ forces	52.9	63.6	41.6	80.0	52.4	69.0	50.9
Sustain theater forces' communications, and computers (C4)	79.0	51.0	20.0	62.5	68.2	62.5	58.8
Conduct force development	41.8	26.8	50.0	53.9	42.2	48.6	42.8
Host-nation security	22.5	37.2	33.3	50.0	44.0	28.6	14.3
Counter or manage deterrence of CBRNE weapons, or operate in a CBRNE environment	75.9	58.4	71.5	75.0	42.0	50.0	52.1
Deploy and maneuver forces	45.0	60.0	66.7	85.7	56.4	72.0	42.9
Safety	41.7	37.8	33.4	16.7	39.4	22.2	37.8
Legal affairs	18.9	39.4	25.1	33.4	38.1	41.1	38.2
Targeting of enemy information systems	43.0	52.8	0.0	50.0	52.0	50.0	54.4
Engineering	57.9	58.9	28.6	14.3	55.8	54.2	51.3
Coordinate counterproliferation in theater	45.0	9.1	0.0	33.3	41.6	50.0	36.4
Conduct mobilization	35.4	18.8	62.5	0.0	42.4	59.0	47.2
Civil affairs and psychological operations	25.9	25.0	0.0	18.2	49.4	44.4	55.5
Mapping, charting, and geodesy	25.7	56.7	50.0	0.0	20.5	33.4	38.3
Maintenance	36.4	42.9	45.5	16.7	36.4	16.7	17.6
Employ firepower or other assets	32.7	16.7	76.9	50.0	18.9	12.5	61.6
Inspector General activities	57.2	66.7	0.0	100.0	52.5	90.0	42.8
Law enforcement	21.1	22.2	37.5	0.0	30.4	0.0	10.0
Medical/health services	53.8	80.0	63.8	20.0	44.8	50.0	72.0
Industrial management	36.9	50.0	16.7	83.4	25.0	0.0	36.4
Combat engineering	20.0	100.0	0.0	0.0	48.0	37.5	0.0

Figure 5.9
Percentage of Respondents Who Consider Tasks that They Perform to Be of Primary/Vital Importance to Their Job

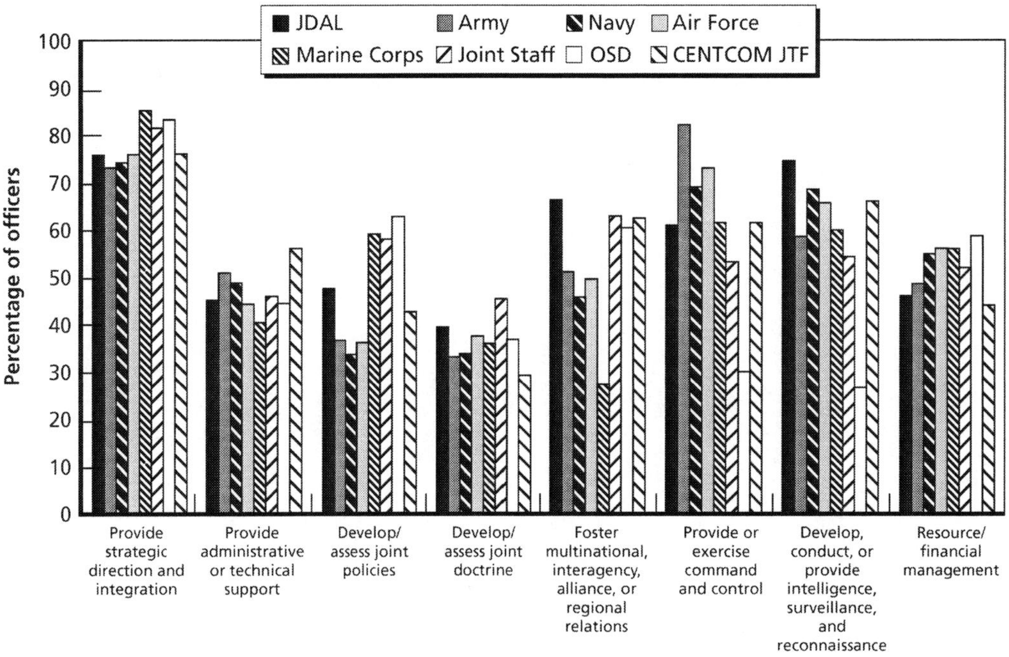

RAND *TR349-5.9a*

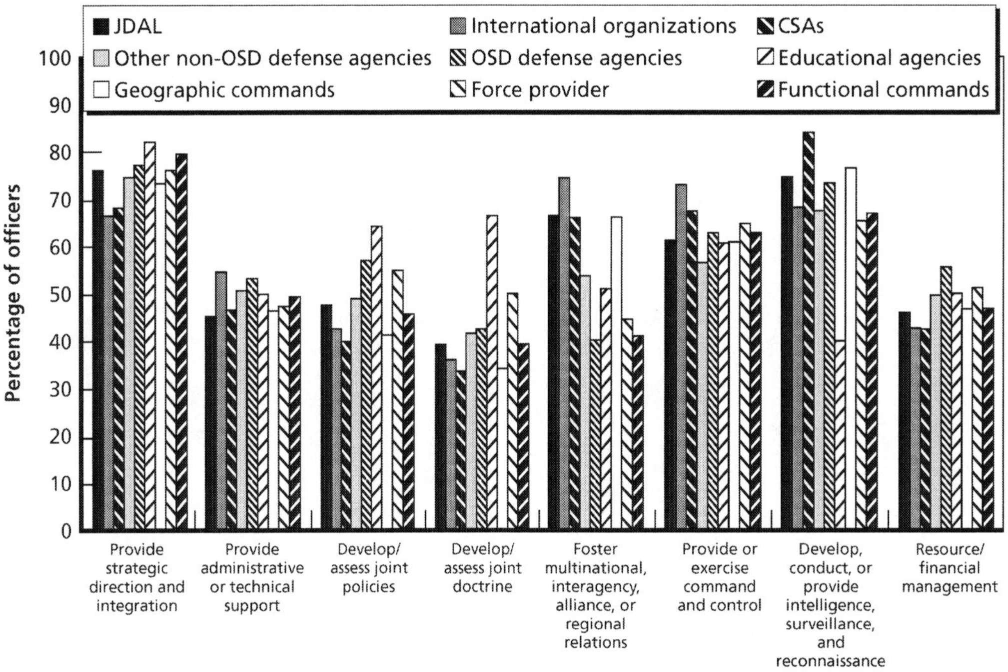

RAND *TR349-5.9b*

- The services ranked the task of providing or exercising command and control particularly high while OSD staff ranked it quite low.
- With the exception of OSD staff and staff in educational agencies, intelligence and surveillance activities were ranked high by most officers, regardless of where the officers were assigned.
- Ratings on the importance of resource/financial management tasks were fairly consistent across organizations.
- In the case of a few specialized agencies (not shown in the tables), all officers performing certain tasks (e.g., civil affairs and psychological operations, and mapping, charting, and geodesy) considered those tasks to be vitally important.

Level of Responsibility for Tasks

For each task respondents performed, they were asked to indicate the level of responsibility for the task on a scale of 1 to 4, where 1 is minimal responsibility, 2 is equally shared responsibility, 3 is primary responsibility, and 4 is sole responsibility. We combined the latter two categories and examined the percentage of officers reporting primary or sole responsibility for a task. It quickly became clear that differences among officers in the level-of-responsibility ratings were being driven largely by the paygrade of the billet. Those in higher-graded billets naturally had greater responsibility for a task. To illustrate this finding, we once again selected the eight top-ranked tasks and examined the percentages of officers, by billet paygrade, who reported that they had primary or sole responsibility for the tasks. The results are shown in Figure 5.10.

Figure 5.10
Percentage of Officers Reporting that They Have Primary/Sole Responsibility for a Task, by Billet Paygrade

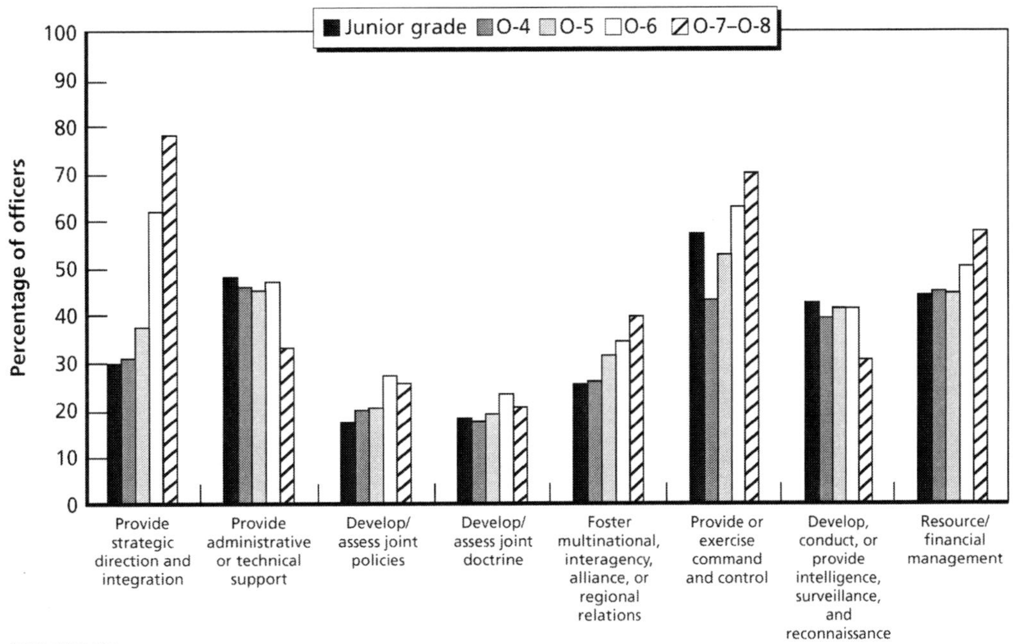

RAND *TR349-5.10*

In general, we find that there is a linear relationship between billet paygrade and higher levels of primary responsibility for tasks that are not primarily operational in nature (such as providing administrative or technical support or conducting intelligence activities). For example, roughly 60 percent of those assigned to O-6 and 80 percent of those in flag officer billets reported primary responsibility for providing strategic direction and integration, compared with about 30 percent of O-4s. The low percentages of higher-grade officers reporting primary responsibility with respect to developing/assessing joint policies and joint doctrine can be partly explained by the fact that these are *joint* policies and, as such, presumably require coordination and cooperation across several agencies and services.

Summary

Table 5.7 shows how billets in various organizations rank on several indicators. This is simply an illustrative exercise and not meant to be exhaustive or definitive. It is useful to consider these kinds of rankings when trying to determine the degree of similarity between billets in different organizations in terms of job characteristics or whether there are some attributes that appear to define "jointness." Based on this set of indicators, JDAL and non-JDAL billets in external organizations rank either first or second, while internal service billets rank third. When we examine major billet organizations, the billets assigned to the Joint Staff, OSD staff, educational agencies, force provider, and the functional commands rank higher on these measures of "jointness" for the most part than the other major billet organizations. The rankings of the other organizations are more mixed, with some ranking high on the number of "highly joint" tasks performed, and others ranking high on the percentage of billets described as "strategic." Lastly, the Marine Corps ranks higher than the other services on the indicators for the four "highly joint" tasks.

Table 5.7
Rankings of Billet Categories Based on Selected Indicators Relating to Tasks Performed

	Rank Based on Percentage of Billets Described as "Strategic"	Rank Based on Percentage of Officers Focused on Defense-Wide Issues	Rank Based on Percentage of Officers Providing Strategic Direction and Integration	Rank Based on Percentage of Officers Developing or Assessing Joint Policies	Rank Based on Percentage of Officers Developing or Assessing Joint Doctrine	Rank Based on Percentage of Officers Fostering Multinational, Interagency, Alliance, or Regional Relations	Rank Based on Percentage of Officers Performing Three or More "Highly Joint" Tasks
JDAL Category							
JDAL billets	1	2	1	1	1	1	1
Non-JDAL billets in external organizations	2	1	2	2	2	2	2
Service-nominated billets	3	3	3	3	3	3	3
Major Billet Organization							
Joint Staff	2	2	2	1	2	8	1
OSD	1	1	1	2	8	6	5
CENTCOM JTF	9	13	12	12	15	3	12
U.S. Army	15	11	15	15	13	11	14
U.S. Navy	14	10	14	14	14	12	15
U.S. Air Force	13	9	13	13	12	13	13
U.S. Marine Corps	10	15	5	6	4	9	8
International organizations	12	12	10	7	7	1	6
CSAs	5	3	11	11	11	5	11
Other Non-OSD defense agencies	4	26	6	10	9	4	9
OSD defense agencies	6	24	4	9	10	14	5
Educational agencies	7	15	9	4	3	7	2
Geographic commands	8	14	7	8	6	2	7
Force provider	11	7	8	5	1	10	4
Functional commands	3	8	3	3	5	15	3

NOTE: Major billet organizations with the same percentages of officers meeting the specific criteria received the same rank.

CHAPTER SIX

Knowledge and Experience Required or Helpful for Job Performance and Experience Gained Through Billet Assignments

To gain a better understanding of officers' job functions, the JOM survey asked respondents about the types of job knowledge and experience that were either helpful for or required by their billets. In parallel, respondents were also asked about the knowledge and experience gained through their billet assignments.

Specific Types of Knowledge

Respondents were asked detailed questions regarding 65 specific types of job knowledge. The types of knowledge fell into 16 broad categories, shown in Table 6.1. The specific types of knowledge that constitute each general-knowledge category are listed in Appendix C.

Usefulness of Specific Types of Knowledge

Within each general-knowledge category, the specific types of knowledge numbered from as few as two to as many as seven. For each item, respondents were asked whether knowledge in that specific area was *required* or *helpful* to someone occupying that billet. Because some types of knowledge were neither required nor helpful for some positions, and respondents

Table 6.1
Categorization of Job Knowledge Items on the JOM Survey

National Military Capabilities, Organization, and Command Structure
National Military Strategy
National Security Strategy
National Security Policy Process
National Planning Systems and Processes
Command, Control, Communications, Computers, Intelligence, Surveillance, and Reconnaissance
Theater Strategy and Campaigning
Geo-Strategic Context
Instruments of National Power
Joint Operation Art
Joint Warfare Fundamentals
Joint Campaigning
Joint Doctrine
Joint and Multinational Forces at the Operational Level of War
Joint Planning and Execution Processes
Other Knowledge Types[a]

[a] Includes knowledge of Inspector General activities and other legal or investigative activities, special operations and operations other than war, manpower and training issues, research and development, medical or health knowledge, acquisition, and joint program management.

did not answer questions regarding those items, we assigned a value of "neither" for missing responses. However, if a respondent skipped the entire set of 65 items, his or her responses were set to "missing" for all items.

Because the questions generated a large amount of data, this discussion focuses on only the 16 broad categories of knowledge shown in Table 6.1. (For further details, see Appendix C.) We averaged over the set of items in each knowledge category by assigning each item equal weight. For purposes of job evaluation, it may be useful to assign differing sets of weights to items, but such an analysis is beyond the scope of this study.

Table 6.2 lists the percentage of officers reporting that particular types of knowledge were required for their billet and the percentage of officers reporting that certain types of knowledge were required *or* helpful for their billet. Table 6.2 also shows the relative rankings of those types of knowledge based on the replies. The two sets of rankings ("required" and "required or helpful") are in general agreement.

There are observable differences in the responses for the various categories of knowledge. Knowledge of National Military Capabilities, Organization, and Command Structure ranks first on the list, with more than 40 percent of officers reporting that such knowledge was required to carry out their assignments and more than 80 percent indicating that such knowledge was required or helpful. Knowledge of the National Security Policy Process was last on the list, with only 13 percent reporting that such knowledge was required for their job and 61 percent reporting that it was required or helpful.

Table 6.2
Percentage of Respondents Reporting that Specific Types of Knowledge Are Required or Helpful for Their Jobs

Knowledge Categories	Percentage Reporting that Knowledge Is "Required" for Their Job	Rank Order	Percentage Reporting that Knowledge Is "Required or Helpful" for Their Job	Rank Order
National Military Capabilities, Organization, and Command Structure	40.7	1	82.7	1
National Military Strategy	25.8	4	71.6	2
National Security Strategy	20.0	11	66.2	7
National Security Policy Process	12.6	16	60.7	16
National Planning Systems and Processes	19.5	14	63.8	10
Command, Control, Communications, Computers, Intelligence, Surveillance, and Reconnaissance	24.9	6	67.4	5
Theater Strategy and Campaigning	27.5	3	67.1	6
Geo-Strategic Context	22.9	8	66.1	9
Instruments of National Power	19.8	12	62.3	13
Joint Operation Art	28.9	2	67.5	4
Joint Warfare Fundamentals	23.9	7	66.2	7
Joint Campaigning	22.8	9	62.8	12
Joint Doctrine	25.7	5	68.2	3
Joint and Multinational Forces at the Operational Level of War	21.2	10	61.9	14
Joint Planning and Execution Processes	19.8	13	61.4	15
Other Knowledge Types	17.8	15	63.6	11

NOTES: Types of knowledge with equal percentages of respondents reporting that the knowledge was required or helpful received the same ranking. The percentages shown here represent the average response rate for the individual items constituting each general category. See Table C.1 for those specific items.

About one-quarter of the officers indicated that knowledge of three areas—National Military Strategy, C4ISR, Theater Strategy and Campaigning, Joint Operation Art, and Joint Doctrine—were required for their jobs. This number goes up to between 67 and 72 percent when we add in the percentage reporting that knowledge in these areas was helpful. We find that with two exceptions—National Military Capabilities, Organization, and Command Structure and National Military Strategy—about 30–40 percent of officers indicated that the types of knowledge were neither required nor helpful for their assignments. There were no open-ended questions on what other types of knowledge might be required or helpful for surveyed positions. Such questions might have identified additional types of knowledge, which could have been useful information for developers of JPME courses.

Table 6.3 presents responses by JDAL category. Officers occupying JDAL billets were more likely than officers in non-JDAL billets or in service-nominated billets to respond that the 65 specific dimensions constituting the 16 general categories were required for or helpful to job performance; on average across all items, 78 percent of officers in JDAL billets responded that the 16 knowledge types were required or useful. Positive responses from officers in JDAL billets ranged from 69 percent to 92 percent across the 16 knowledge categories. In contrast, officers in internal service billets (i.e., service-nominated billets) were, on average, the least likely to indicate that the specific knowledge types were required for or helpful to their positions; those officers had an average positive response of 59 percent across all items. Officers in non-JDAL billets in external organizations generally tended to fall in the middle.

Despite the differences in percentages among the three groups, the rank order of the knowledge categories according to the percentage replying in the affirmative is fairly consistent across the groups.[1] For instance, National Military Capabilities, Organization, and Command Structure ranked first across all three categories of billets.

If we disaggregate the data to examine responses on the individual survey items, we find that officers in JDAL billets ranked some specific types of knowledge very high. For example, 96 percent reported that knowledge of the Roles, Relationships, and Functions of the National Command Authority (NCA), JCS, COCOMs, National Security Council (NSC), joint forces commander (JFC), CJCS, or the Interagency Process (under the general category National Military Capabilities, Organization, and Command Structure) was very useful; this also was the highest percentage for any item across all three groups of officers. Conversely, knowledge about Medical or Health Services (under the Other Knowledge Types general category) was ranked relatively low, with only 51–62 percent of officers across the three categories indicating that such knowledge was required or helpful for job performance.

Overall, it appears that occupants of JDAL billets believe that their positions require a broader set of knowledge and more specific types of knowledge than the types of knowledge occupants of non-JDAL billets, especially those in service-nominated billets, believe are required for their jobs.

---

[1] The correlation coefficient between the rank orderings was 0.80, which indicates a moderately high degree of agreement between the two.

Table 6.3
Percentage of Respondents Reporting that Specific Types of Knowledge Are Required or Helpful, by JDAL Category

Knowledge Categories	JDAL Billets	Non-JDAL Billets in External Organizations	Service-Nominated Billets
National Military Capabilities, Organization, and Command Structure	91.6	82.3	77.2
National Military Strategy	83.5	70.3	64.8
National Security Strategy	78.0	66.4	58.9
National Security Policy Process	74.3	61.6	52.0
National Planning Systems and Processes	77.3	64.7	55.1
Command, Control, Communications, Computers, Intelligence, Surveillance, and Reconnaissance	75.1	66.9	62.9
Theater Strategy and Campaigning	79.8	66.5	59.5
Geo-Strategic Context	79.0	67.1	57.8
Instruments of National Power	76.4	61.7	53.8
Joint Operation Art	79.2	67.1	60.5
Joint Warfare Fundamentals	77.1	65.6	59.8
Joint Campaigning	73.8	63.3	55.8
Joint Doctrine	79.6	67.7	61.4
Joint and Multinational Forces at the Operational Level of War	73.0	60.5	55.7
Joint Planning and Execution Processes	73.5	60.8	54.1
Other Knowledge Types	69.2	62.7	60.6
Overall minimum	69.2	60.5	52.0
Overall average	77.5	66.0	59.4
Overall maximum	91.6	82.3	77.2

NOTES: Types of knowledge with equal percentages of respondents reporting that the knowledge was required or helpful received the same ranking. The percentages shown here represent the average response rate for the individual items constituting each general category. See Table C.1 for those specific items.

Perceptions of the helpfulness of the 16 broad types of knowledge and whether those types of knowledge are required vary sharply by billeted grade (see Table 6.4). Occupants of junior-grade billets (O-1–O-3) are much less likely to view the knowledge categories as being required or helpful than are the occupants of flag-officer billets (O-7–O-10), with those occupying the O-4–O-6 billets falling somewhere in between. For example, about 71 percent of junior-grade billet officers rated knowledge of National Military Capabilities, Organization, and Command Structure as being required or helpful—the highest percentage among these officers across all knowledge categories. In contrast, the *lowest* percentage of officers in flag billets reporting that a specific type of knowledge was helpful was approximately 79 percent (for knowledge of the National Security Policy Process). On average, across the 16 broad knowledge categories, about 54 percent of the occupants of junior-grade billets rated the knowledge types as being required or helpful, compared with 69 percent of those in mid-level and 84 percent of those in flag officer billets.

Knowledge Gained Through Job Assignment

A separate set of questions asked respondents to indicate whether serving in their positions would allow them to gain proficiency in the various types of knowledge. We find marked

Table 6.4
Percentage of Officers Reporting that Specific Types of Knowledge Are Required or Helpful, by Billet Paygrade

Knowledge Categories	Junior Officer	Mid-Level Officer	Flag Officer
National Military Capabilities, Organization, and Command Structure	71.2	85.8	95.8
National Military Strategy	58.0	75.1	90.4
National Security Strategy	52.7	69.6	87.3
National Security Policy Process	48.1	64.0	78.6
National Planning Systems and Processes	49.7	67.5	83.3
Command, Control, Communications, Computers, Intelligence, Surveillance, and Reconnaissance (C4ISR)	59.3	69.4	80.7
Theater Strategy and Campaigning	54.0	70.6	84.4
Geo-Strategic Context	54.3	69.2	83.2
Instruments of National Power	48.8	65.8	81.2
Joint Operation Art	54.2	71.0	83.6
Joint Warfare Fundamentals	54.3	69.3	82.8
Joint Campaigning	50.6	66.0	81.7
Joint Doctrine	55.1	71.7	84.7
Joint and Multinational Forces at the Operational Level of War	49.9	65.0	81.1
Joint Planning and Execution Processes	49.1	64.5	79.3
Other Knowledge Types	54.8	65.6	83.4
Overall minimum	48.1	64.0	78.6
Overall average	54.0	69.4	83.8
Overall maximum	71.2	85.8	95.8

NOTE: The percentages shown here represent the average response rate for the individual items constituting each general category. See Table C.1 for those specific items.

differences across the 16 knowledge categories with respect to the level of knowledge respondents perceived they would gain through their assignments. Many of these answers are tied to officers' previous responses regarding whether such knowledge was required or helpful for an assignment, so there is a strong correlation between respondents' perception of the proficiency that would be gained in an area and whether that knowledge was thought to be helpful or required. In examining these responses, one should keep in mind that we do not have information on the proficiency/familiarity levels of officers with respect to the given types of knowledge *prior* to the officers' occupying their billets. As a result, one should not interpret the responses as reflecting knowledge gained solely during an assignment.

Table 6.5 shows the percentage of officers who indicated that they are likely to gain proficiency or familiarity with a specific type of knowledge on the job, by knowledge category. Table 6.5 also presents rankings based on two indicators: the mean percentage of respondents who indicated an assignment would lead to proficiency with respect to a certain knowledge category and the mean percentage of respondents who indicated that their assignment would allow them to become either proficient or familiar with a type of knowledge. (Further details are available in Appendix C.)

There is a wider range of responses for the likelihood of attaining proficiency compared with attaining familiarity or proficiency with an area of knowledge, which is not surprising, given that familiarity is easier to attain and respondents may differ in their definition of proficiency. There is some indication that job assignments do not provide the specific types of knowledge about which the survey asked, since more than one-third of respondents on average indicated that their assignment would provide neither familiarity nor proficiency in the knowledge areas surveyed.[2] If these specific types of knowledge are thought to be important to joint duty assignments, then formal training or education may be needed to supplement on-the-job training.

Levels of proficiency with respect to both general and specific knowledge types vary across billets. For example, 38 percent of surveyed officers indicated that their billets would allow them to become proficient in National Military Capabilities, Organization, and Command Structure, and about 80 percent reported that they would gain either proficiency or familiarity in that area through their current assignment. Conversely, only 12 percent indicated that they would achieve proficiency and 58 percent indicated that they would gain either proficiency or familiarity with the National Security Policy Process during their assignment. The two relative rankings in Table 6.5 are in general agreement.

When we examine differences in responses by the JDAL status of the billets, we find that officers occupying JDAL billets were more likely on average (75 percent) to respond that

Table 6.5
Percentage of Officers Reporting that Their Assignment Will Lead to Proficiency or Familiarity with Specific Types of Knowledge

Knowledge Category	Percentage Reporting that Assignment Will Lead to Proficiency	Rank	Percentage Reporting that Assignment Will Lead to Proficiency or Familiarity	Rank
National Military Capabilities, Organization and Command Structure	38.0	1	79.9	1
National Military Strategy	23.8	4	69.1	2
National Security Strategy	19.1	11	64.0	7
National Security Policy Process	11.8	16	58.3	15
National Planning Systems and Processes	18.0	13	61.4	10
Command, Control, Communications, Computers, Intelligence, Surveillance, and Reconnaissance	22.3	5	65.0	3
Theater Strategy and Campaigning	24.8	3	64.5	6
Geo-Strategic Context	21.6	8	63.7	8
Instruments of National Power	18.9	12	59.8	13
Joint Operation Art	25.8	2	64.9	4
Joint Warfare Fundamentals	22.1	6	63.6	9
Joint Campaigning	21.2	9	60.3	12
Joint Doctrine	22.1	7	64.8	5
Joint and Multinational Forces at the Operational Level of War	20.0	10	59.1	14
Joint Planning and Execution Processes	17.7	14	58.3	15
Other Knowledge Types	16.9	15	61.3	11
Overall average	21.5		63.6	

NOTE: The percentages shown here represent the average response rate for the individual items constituting each general category. See Table C.1 for those specific items.

[2] It may be that respondents who were already proficient or familiar with such knowledge answered no to this question because they did not gain such knowledge during the assignment.

their assignment would lead to proficiency or familiarity with the 65 specific dimensions of knowledge that constitute the 16 general categories than were officers in the other two types of billets (at 57 and 63 percent on average) (see Table 6.6). Officers across the board were much less likely to report gaining familiarity with Joint and Multinational Forces at the Operational Level of War, Joint Planning and Execution Processes, and Other Knowledge Types than the other categories of knowledge.

There was a large (15–23 percentage point) difference between JDAL billets and other billets in the likelihood of gaining familiarity with Instruments of National Power. Overall, the percentages for non-JDAL billets and service-nominated billets look similar, but they differ from the JDAL billet percentages with respect to types of knowledge gained on the job.

One noteworthy finding from the more detailed examination (found in Appendix C) is the very high percentage of officers (75 percent) in service-nominated billets reporting that their assignment would lead to familiarity with the roles, relationships, and functions of the NCA, JCS, COCOMs, NSC, JFC, CJCS, or the interagency process, the highest percentage for any specific knowledge type among all service-nominated billets.

Table 6.6
Percentage of Officers Reporting that Their Assignment Will Lead to Proficiency or Familiarity with Specific Types of Knowledge, by JDAL Category

Knowledge Categories	JDAL Billets	Non-JDAL Billets in External Organizations	Service-Nominated Billets
National Military Capabilities, Organization, and Command Structure	89.6	79.7	74.1
National Military Strategy	81.4	67.7	62.1
National Security Strategy	75.8	64.0	56.7
National Security Policy Process	71.8	58.8	49.8
National Planning Systems and Processes	74.9	62.0	52.6
Command, Control, Communications, Computers, Intel, Surveillance, and Reconnaissance	73.0	64.2	60.4
Theater Strategy and Campaigning	77.5	63.7	56.8
Geo-Strategic Context	76.8	64.6	55.2
Instruments of National Power	74.0	59.1	51.3
Joint Operation Art	76.8	64.2	57.8
Joint Warfare Fundamentals	74.7	63.0	57.0
Joint Campaigning	71.4	60.1	53.5
Joint Doctrine	76.3	64.1	58.0
Joint and Multinational Forces at the Operational Level of War	70.4	57.6	52.8
Joint Planning and Execution Processes	70.4	57.7	51.1
Other Knowledge Types	67.2	60.2	58.0
Overall minimum	67.2	57.6	49.8
Overall average	75.1	63.2	56.7
Overall maximum	89.6	79.7	74.1

NOTE: The percentages shown here represent the average response rate for the individual items constituting each general category. See Table C.1 for those specific items.

Perceptions about the efficacy of job functions in transmitting proficiency or familiarity with specific types of knowledge vary significantly by billet paygrade, as shown in Table 6.7. Compared with occupants of junior-grade billets, flag officers were far more likely to report gaining proficiency with knowledge areas through their assignments (81 percent compared with 51 percent). There is a marked difference between mid-level officers and flag officers in their likelihood of becoming proficient or familiar with any of the knowledge areas.

Linking the Types of Knowledge Necessary for a Billet and Gained Through a Billet Assignment to Performance of "Highly Joint" Tasks

The types of knowledge needed for a billet are likely to vary by the tasks performed in the billet. We examined the responses of officers who performed the four "highly joint" tasks identified in Chapter Five—perform strategic direction and integration; develop/assess joint policies; develop/assess joint doctrine; and foster multinational, interagency, or regional relations—to see whether and how their responses differed.

Figure 6.1 shows (1) the mean percentage of officers (averaged across responses on the 16 categories of knowledge) who reported that the 16 types of knowledge were required or helpful for their billet assignment and (2) the mean percentage of officers who reported that they would gain proficiency or familiarity in these areas by the end of their assignment. Officers were grouped by JDAL category.

Table 6.7
Percentage of Officers Reporting that Their Assignment Will Lead to Proficiency or Familiarity with Specific Types of Knowledge, by Billet Paygrade

Knowledge Categories	Junior Officer	Mid-Level Officer	Flag Officer
National Military Capabilities, Organization and Command Structure	67.4	83.4	93.4
National Military Strategy	55.1	72.7	87.9
National Security Strategy	50.2	67.5	84.8
National Security Policy Process	45.5	61.7	76.9
National Planning Systems and Processes	47.2	65.0	80.4
Command, Control, Communications, Computers, Intel, Surveillance, and Reconnaissance	56.3	67.2	79.0
Theater Strategy and Campaigning	51.1	68.1	80.6
Geo-Strategic Context	51.1	67.0	81.3
Instruments of National Power	46.1	63.4	78.2
Joint Operation Art	51.0	68.7	80.9
Joint Warfare Fundamentals	51.0	66.9	80.4
Joint Campaigning	47.5	63.6	79.1
Joint Doctrine	51.1	68.5	82.1
Joint and Multinational Forces at the Operational Level of War	46.3	62.4	77.9
Joint Planning and Execution Processes	45.8	61.6	75.8
Other Knowledge Types	51.7	63.4	81.2
Overall minimum	45.5	61.6	75.8
Overall average	50.9	66.9	81.2
Overall maximum	67.4	83.4	93.4

NOTE: The percentages shown here represent the average response rate for the individual items constituting each general category. See Table C.1 for those specific items.

Figure 6.1
Percentage of Officers Performing "Highly Joint" Tasks Who Reported that All Types of Knowledge Are Required or Helpful and that They Would Become Proficient or Familiar with That Knowledge by the End of Their Assignment, by JDAL Category

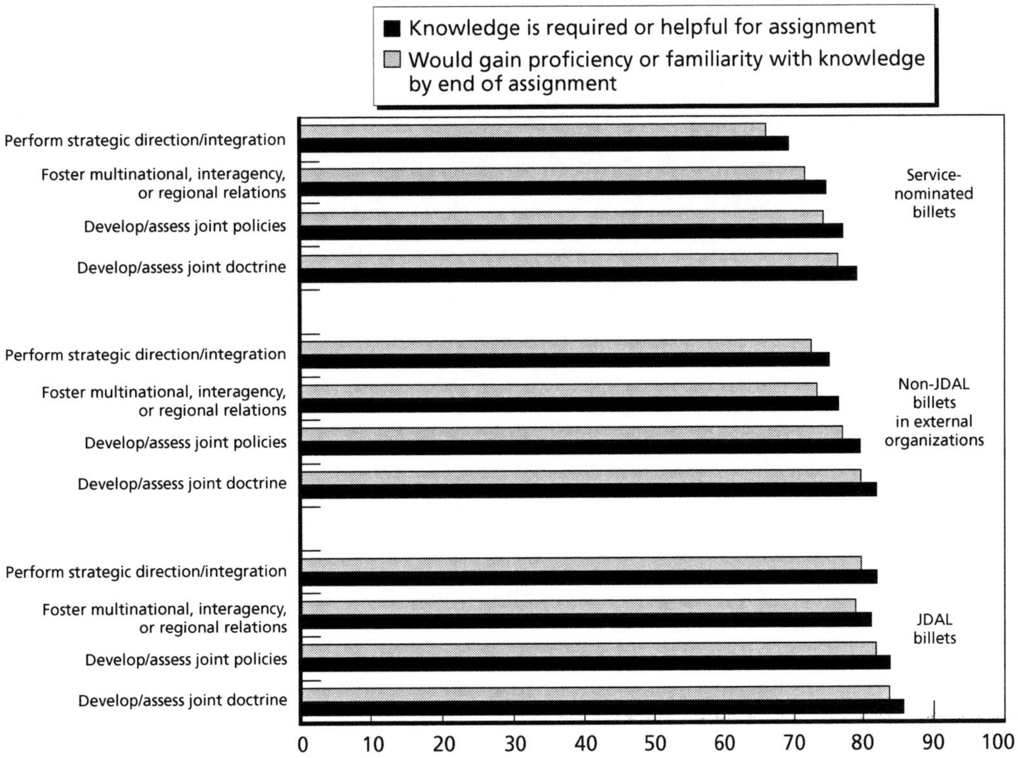

RAND *TR349-6.1*

Between 75 and 85 percent of officers performing these highly joint tasks identified all 16 types of knowledge as required or helpful for their assignment, and most of them reported that they would become proficient or familiar with these areas by the end of their assignment. As is evident, compared with officers assigned to JDAL billets, officers in service-nominated billets, particularly those providing strategic direction and integration, were somewhat less likely to report that knowledge of the 16 general types of knowledge enumerated in the survey was useful for their billet or that they would gain proficiency or familiarity with those areas in the course of their assignment. If we look at individual knowledge categories (not shown in the figure), there appears to be little difference among the officers across the three billet categories with respect to ranking some types of knowledge higher or lower. This suggests that, at least for the highly joint tasks, a broad range of knowledge is required or useful for carrying out these tasks and concomitantly, officers believe that they are gaining knowledge through their assignments.[3]

Respondents also were asked whether certain types of training, experience, or education would be helpful to their job performance. Their responses to this question are discussed in the next section.

---

[3] It would be useful to compare this pattern of responses for other sets of tasks to determine whether those tasks require a different set of knowledge in order to perform them. Such a determination may have implications for structuring joint professional education and training for different groups of billets.

Joint Training and Education, Experience, and Expertise

A set of survey questions attempted to measure officers' opinions about the kinds of education, experience, skills, and competencies that were valuable to them in their assignments. These questions included those on (1) the value of JPME I and JPME II, other joint training, and prior joint experience; (2) the extent to which officers drew upon their primary specialty and knowledge of their own service's capabilities; and (3) rankings of the skill, expertise, and experience that were most important to them in their assignment and the skill, expertise, and experience that would be most important for their successors to possess. The response categories and the way in which the questions were worded differed, so it is not always possible to directly compare the findings reported in the following sections.

Value of JPME I, JPME II, Other Joint Training and Education, and Prior Joint Experience

The majority of officers responded that JPME I, JPME II, other joint training or education (other than JPME), and prior experience in a joint environment were required or desired to perform their duties successfully (see Figures 6.2 and 6.3).[4] Officers in internal service billets were less likely to respond as such, but even among those officers, between 70 and 80 percent believed such training, education, and experience would be required or desired.

Figure 6.2
Percentage of Officers Reporting that Joint Education, Training, and Experience Are Required or Desired, by JDAL Category

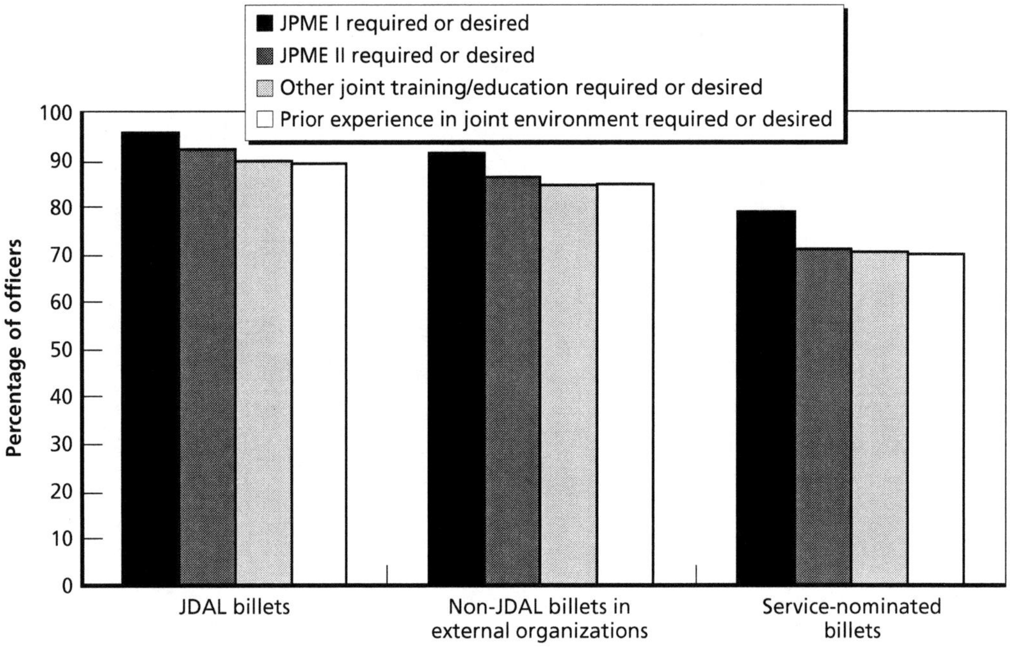

[4] Large percentages of officers reported that they had no experience with JPME I or JPME II. For example, more than 40 percent of officers in non-JDAL billets in external organizations indicated that they had no experience with JPME I, as did 48 percent of officers in internal service billets. The percentages were even higher for those reporting no experience with JPME II (52 percent of those in non-JDAL billets in external organizations and 59 percent of those in internal service billets). We restricted the responses to those with experience with JPME when calculating the percentages of those who reported that JPME I or JPME II was required or desired.

Figure 6.3
Percentage of Officers Reporting that Joint Education, Training, and Experience Are Required or Desired, by Major Billet Organization

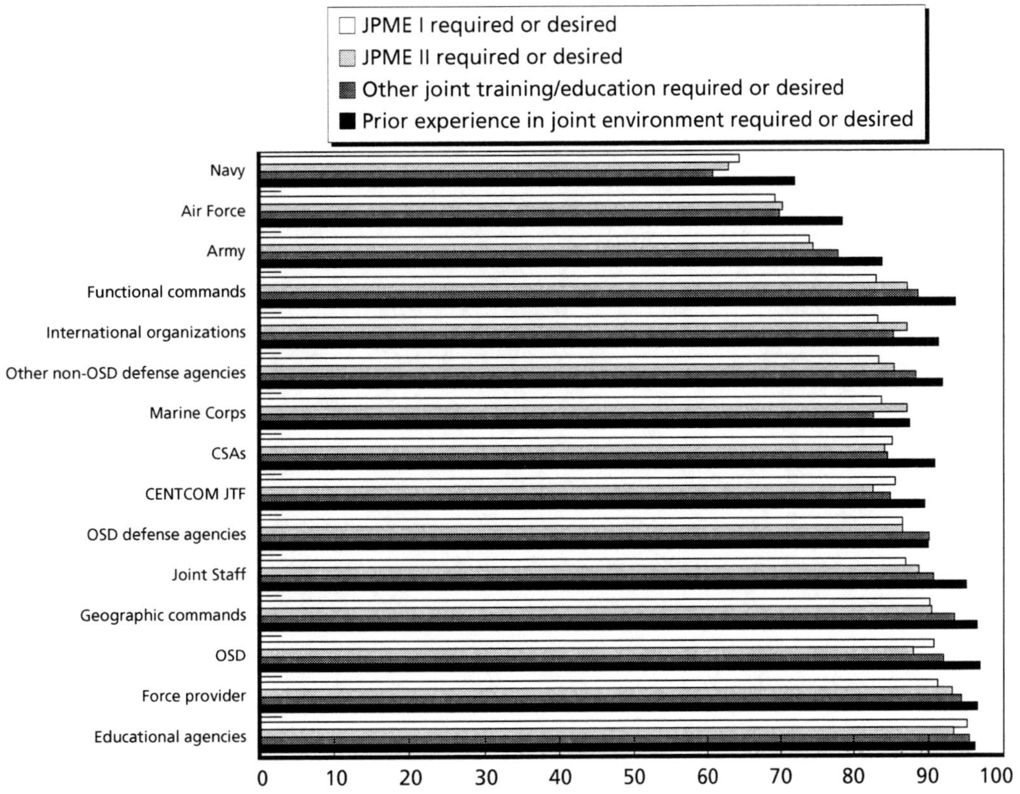

RAND *TR349-6.3*

Among major billet organizations, well over 80 percent of officers in non-service billets and in the Marine Corps billets reported that joint training, education, and experience were required or desired for their assignments.

Importance of Primary Specialty and Knowledge of Service's Capabilities

Not unexpectedly, officers in internal service billets were the most likely to report drawing upon their primary specialty and knowledge of their own service's capabilities to carry out their assignments most of or all of the time (see Figures 6.4 and 6.5). For example, 63 percent of those officers used their knowledge of their service's capabilities most or all of the time in their positions, compared with 30–34 percent of those in non-service billets. Among the services, officers in Marine Corps billets overwhelmingly responded that they relied on their knowledge of their service's capabilities to carry out their jobs. Officers assigned to the Joint Staff or educational agencies were the least likely to draw upon either their primary specialty or knowledge of own service's capabilities.

General and flag officers were much more likely to report drawing on their knowledge of their service's capabilities (71 percent) than on their primary specialty (51 percent) most or all of the time. About 53–65 percent of O-6s and below reported drawing on their primary specialty most or all of the time; the comparable numbers for those who depend on knowledge of their service's capabilities were 44–49 percent.

Figure 6.4
Percentage of Officers Reporting that They Draw on Their Primary Specialty and Knowledge of Their Own Service's Capabilities Most or All of the Time, by JDAL Category

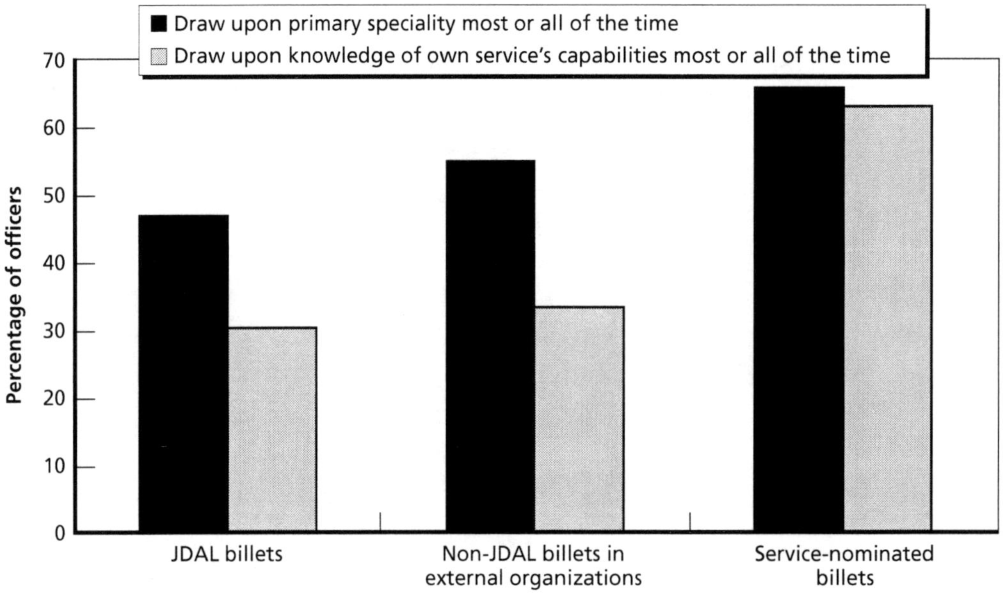

RAND *TR349-6.4*

Figure 6.5
Percentage of Officers Reporting that They Draw on Their Primary Specialty and Knowledge of Their Own Service's Capabilities Most or All of the Time, by Major Billet Organization

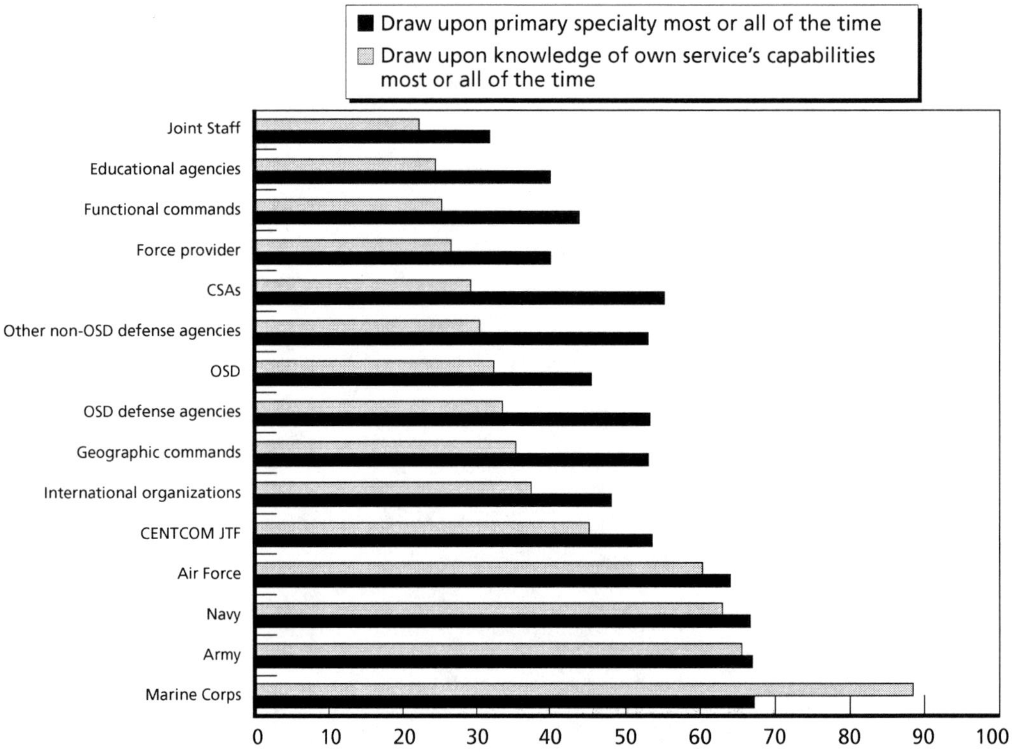

RAND *TR349-6.5*

Most Important Skill/Expertise/Experience for Job

Respondents were asked to rate the most important skill/expertise/experience from a list of five items for both themselves and their successors. Table 6.8 shows the percentage of respondents naming each of the five items, grouped by JDAL billet category. In the following discussion, we combine the two items relating to functional expertise; therefore, the comparisons are being made across four items.

About 35 percent of officers in JDAL billets and 43 percent of officers in non-JDAL, non-service billets rated the combined functional expertise areas as the most important expertise/skill in their assignment, followed by service core competencies (20–21 percent). Between 15 and 17 percent ranked prior joint experience first. Specialized training and orientation in joint matters was selected as being important by only 15 percent of JDAL officers and 9 percent of officers in non-JDAL billets in external organizations.

For officers in service-nominated billets, about 38 percent ranked service core competencies as most important, and an equal percentage rated the two combined functional expertise areas as most important. Only 7 percent ranked prior joint experience or specialized training in joint matters as being most important.

The rankings were consistent across the two question categories. The skills and expertise ranked highly by respondents for their own job performance were also ranked highly as skills and expertise that the respondents' successors would need to carry out the duties of their assignments successfully.

Table 6.9 shows the rankings of the various types of competencies and experience by officer paygrade for incumbents only.

Table 6.8
Percentage of Officers Selecting a Given Skill/Expertise/Experience as Being the Most Important to Them and Their Successors, by JDAL Category

	JDAL Billets	Non-JDAL Billets in External Organizations	Service-Nominated Billets
Most important to you in your assignment			
Service core competencies	20.1	21.2	37.8
Prior joint experience	17.4	14.7	7.0
Specialized training and orientation in joint matters	14.6	9.1	7.0
Functional expertise other than acquisition/joint program management	29.5	29.2	22.4
Functional expertise in acquisition/joint program management	5.4	13.6	15.7
Other	13.0	12.1	10.1
Most important for your successor to possess			
Service core competencies	18.3	18.3	36.4
Prior joint experience	14.6	12.9	6.0
Specialized training and orientation in joint matters	17.5	11.3	7.5
Functional expertise other than acquisition/joint program management	29.6	29.3	21.8
Functional expertise in acquisition/joint program management	6.0	15.1	17.9
Other	13.9	13.2	10.5

Table 6.9
Percentage of Officers Selecting a Given Skill/Expertise/Experience as Being the Most Important to Them and Their Successors, by Paygrade

	Junior Officers	O-4s	O-5s	O-6s	General/Flag Officers
Most important to you in your assignment					
Service core competencies	32.4	26.2	26.1	24.7	45.9
Prior joint experience	5.8	9.4	13.5	21.5	17.0
Specialized training and orientation in joint matters	9.1	10.9	9.7	6.6	3.1
Functional expertise other than acquisition/joint program management	25.2	29.5	25.2	21.4	15.9
Functional expertise in acquisition/joint program management	15.1	11.9	13.4	12.5	9.5
Other	12.4	12.1	11.8	13.2	8.7
Most important for your successor to possess					
Service core competencies	31.4	26.1	24.6	23.2	46.4
Prior joint experience	5.5	7.2	11.6	19.0	16.9
Specialized training and orientation in joint matters	9.8	13.5	12.4	9.3	2.5
Functional expertise other than acquisition/joint program management	25.2	28.9	25.6	22.2	15.8
Functional expertise in acquisition/joint program management	15.6	12.6	14.2	12.9	9.6
Other	12.6	11.9	11.6	13.4	8.8

General and flag officers rated service core competencies very highly, with 46 percent reporting that those competencies were of greatest importance in their assignments and also the most important for their successors to possess. All other officer grades ranked the two combined areas of functional expertise as the most important knowledge to possess.

Both O-6s and flag officers tended to rate prior joint experience much more highly than did other officers, with 22 percent and 17 percent, respectively, rating this type of experience as the most important for them to have, and 17 and 19 percent, respectively, rating it as the most important for their successors to have. About 12–13 percent of junior and O-4–O-6 officers and 9 percent of flag officers rated "other" competencies/knowledge/experience first in the list, although we do not have information on what those other items might be. Specialized training and orientation in joint matters generally ranked low on the list and was ranked the lowest by senior officers.

Figure 6.6 shows respondents' rankings on the importance of service core competencies, prior joint experience, and functional expertise, by major billet organization.

Not surprisingly, officers in internal service billets (with the exception of the Air Force) ranked service core competencies as the most important to them in their assignments, while officers in non-service billets gave functional expertise top billing. Officers assigned to educational agencies ranked prior joint experience as the most important, while the responses of those assigned to force provider billets were almost evenly split across the three categories.

Figure 6.6
Percentage of Officers Reporting that Service Core Competencies, Prior Joint Experience, or Functional Expertise Was Most Important in Their Assignments, by Major Billet Organization

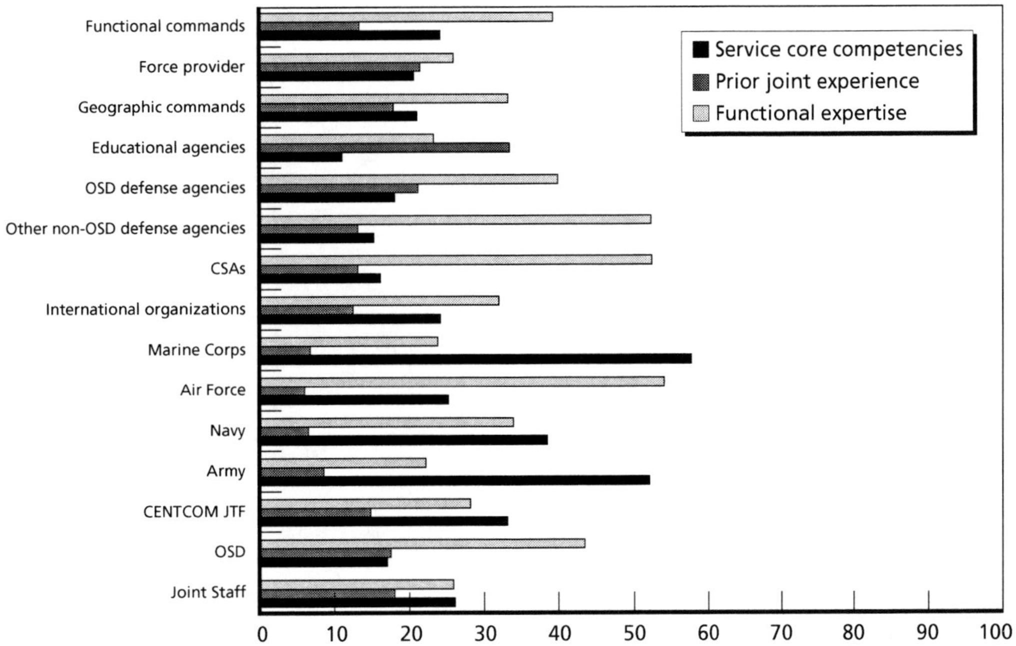

RAND *TR349-6.6*

Types of Experience Provided by a Billet

Respondents were asked about significant experience gained through assignment to the billet in three areas: multiservice, multinational, and interagency. Figure 6.7 shows the percentage of officers in the three JDAL categories who "agreed" or "strongly agreed" that their billet provides them with significant experience in these three areas.

JDAL billets provided the most experience across all three areas. Overall, 87 percent of officers in JDAL billets reported that they gained significant experience in multiservice matters, 65 percent reported gaining significant experience in multinational matters, and 75 percent reported as such for interagency matters. Officers in non-JDAL, non-service billets were much less likely to report gaining experience with multinational matters than to report gaining experience with multiservice or interagency matters. Officers in service-nominated billets were less likely than those in non-service billets to report gaining experience in these areas.

There were some significant differences in the experience afforded officers attached to the various billet organizations, as shown in Figures 6.8 through 6.10.

More than 70 percent of officers in non-service organizations (with one exception) reported that their billets in these organizations provide significant experience in multiservice matters, and this is true of Marine Corps billets as well. This compares with the 46–52 percent of respondents in the Army, Navy, and Air Force billets who replied as such. The differences are more marked if one examines only those who strongly agreed that their billet

Figure 6.7
Percentage of Officers Who Agreed or Strongly Agreed that Their Position Provides Significant Experience in Multiservice, Multinational, or Interagency Matters, by JDAL Category

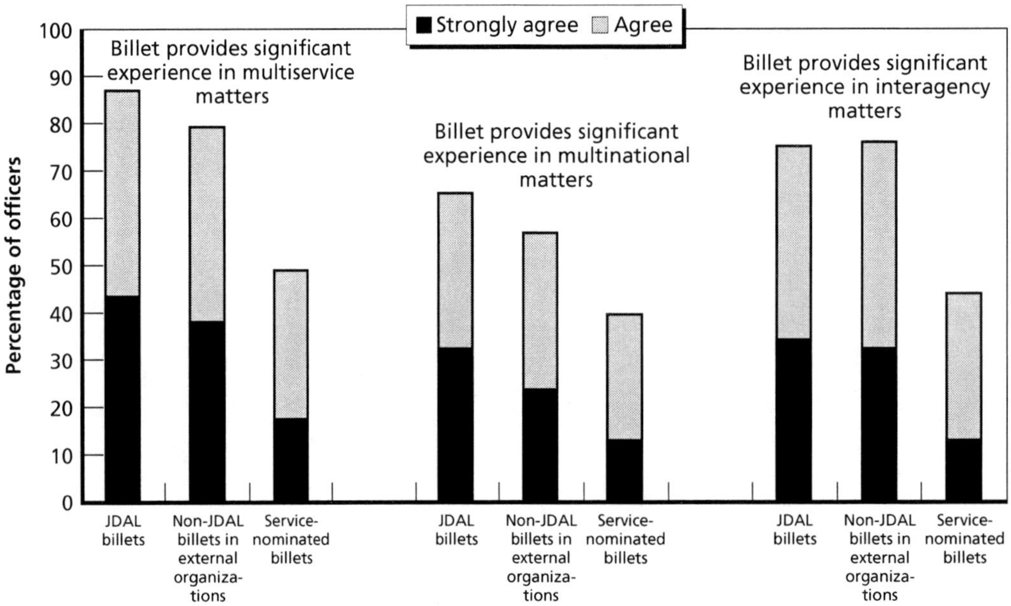

RAND *TR349-6.7*

Figure 6.8
Percentage of Officers Who Agreed or Strongly Agreed that Their Position Provides Significant Experience in Multiservice Matters, by Major Billet Organization

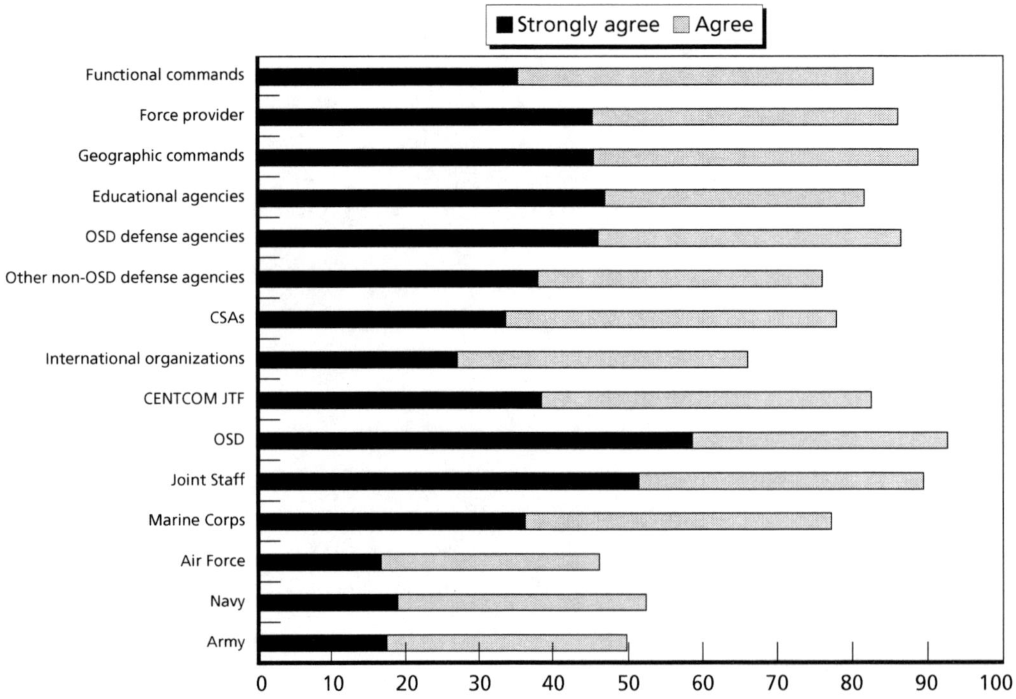

RAND *TR349-6.8*

provides such experience. More than 50 percent of those assigned to the Joint Staff or OSD staff strongly agreed that their billet provides them with significant experience in multiservice matters. Others that ranked high on this indicator were educational agencies, OSD defense agencies, geographic commands, and force provider billets.

Almost all officers assigned to international organizations reported getting significant experience in multinational matters, and 80 percent were in strong agreement with this statement. This percentage was considerably higher than the percentage in this category for any other organization. Other organizations that ranked high on this indicator included CENTCOM JTF, geographic commands, and educational agencies, with roughly 70–80 percent of officers in those organizations agreeing or strongly agreeing that their billets provide significant experience in multinational matters.

With a few exceptions, non-service billets provided significant amounts of experience in interagency matters. Those serving in OSD staff or other agency billets were particularly likely to strongly agree that their billets provided such experience. About 40–45 percent of officers in internal service billets reported gaining such experience.

We were also interested in examining which of the organizations/billets provided experience in more than one of these areas. Figures 6.11 and 6.12 show the percentage of officers who reported getting experience in one, two, or all three areas (multiservice, multinational, and interagency matters) for billets grouped by JDAL category and for billets grouped by major billet organization.

Figure 6.9
Percentage of Officers Who Agreed or Strongly Agreed that Their Position Provides Significant Experience in Multinational Matters, by Major Billet Organization

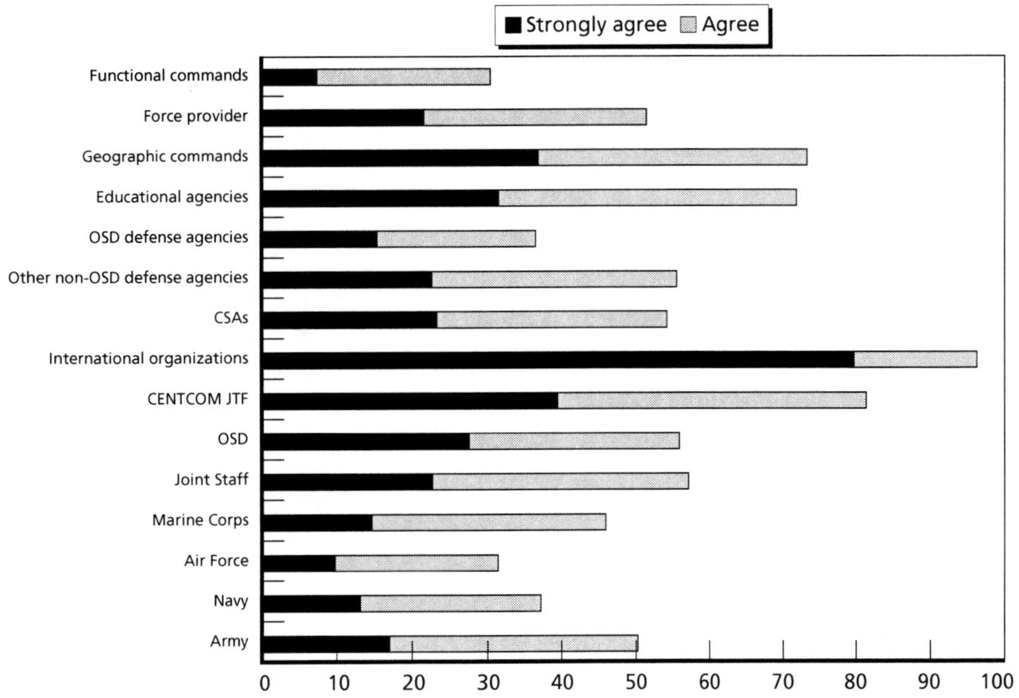

RAND *TR349-6.9*

Figure 6.10
Percentage of Officers Who Agreed or Strongly Agreed that Their Position Provides Significant Experience in Interagency Matters, by Major Billet Organization

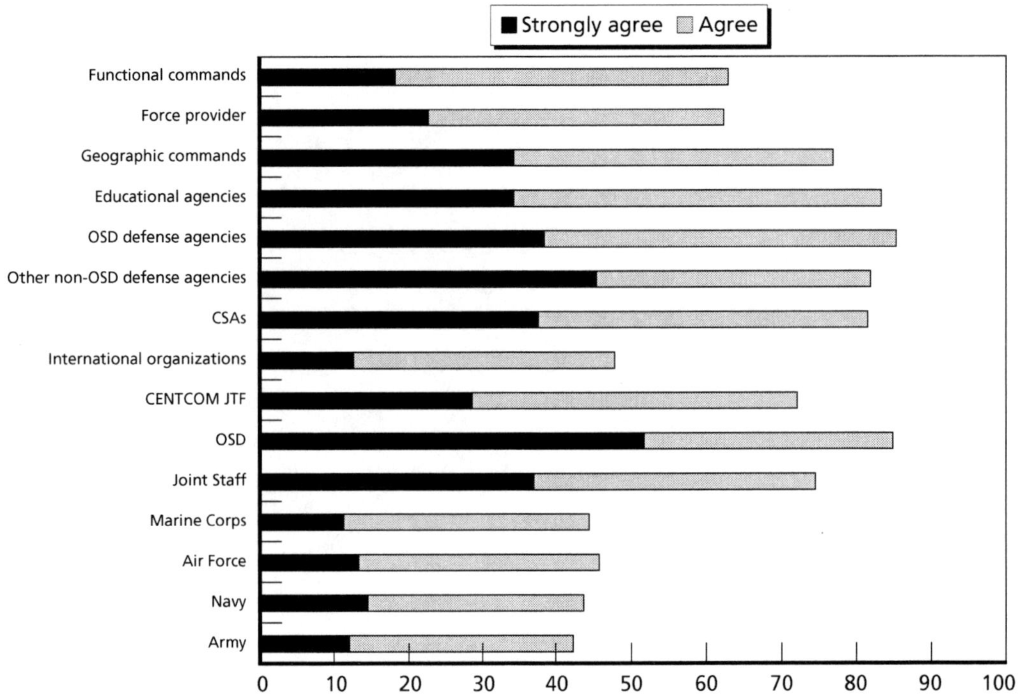

RAND *TR349-6.10*

Ninety percent or more of JDAL and non-JDAL non-service billets provide experience in at least one area, 75 percent of JDAL billets and 80 percent of non-JDAL billets provide experience in at least two areas, and 53 percent of JDAL billets and 47 percent of non-JDAL billets provide experience in all three areas. For internal service billets, the percentages were considerably lower, with 65 percent, 43 percent, and 24 percent, respectively, providing significant experience in at least one, at least two, or all three areas.

Well over 85 percent of billets in all but the services provide significant experience in at least one of these areas, and, with some exceptions, well over 70 percent of those billets provide significant experience in two of the areas. More than half of the billets in the educational agencies, CENTCOM JTF, geographic commands, Joint Staff, and OSD staff provide significant experience in all three areas.

Summary

Overall, officers ranked knowledge of National Military Capabilities, Organization, and Command Structure first among the 16 knowledge categories, with more than 80 percent indicating that such knowledge was required or desired for their assignments. Although knowledge of the National Security Policy Process was last on the list, 61 percent of respondents reported that it was required or helpful. With some exceptions, about 30–40 percent of officers indicated that particular types of knowledge were neither required nor desired for their assignments.

Figure 6.11
Percentage of Officers Who Agreed or Strongly Agreed that Their Position Provides Significant Experience in One, Two, or Three Key Areas, by JDAL Category

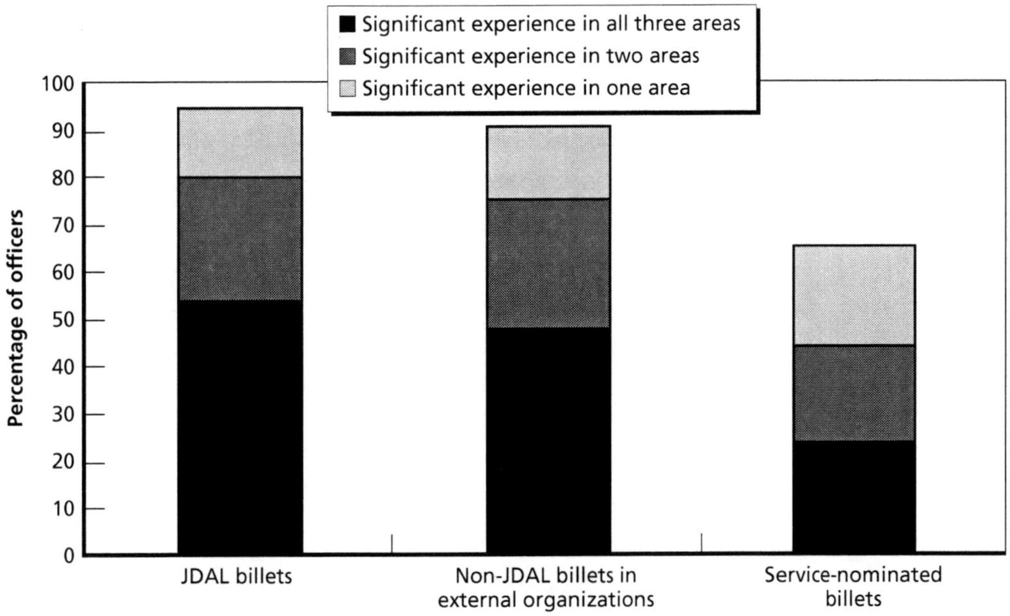

RAND *TR349-6.11*

Officers indicated that some types of knowledge (such as knowledge of National Military Capabilities, Organization, and Command Structure) were more likely to be gained through their assignments than others (such as knowledge of the National Security Policy Process). Overall, more than one-third of officers reported not gaining proficiency or familiarity with any of the different sets of knowledge through their assignments.

Overall, officers in JDAL billets and senior officers were much more likely than those in service-nominated billets and junior officers to rank various types of knowledge as being both required and desired and to report that their assignment would allow them to gain proficiency or familiarity in an area.

In the previous two chapters, we attempted to rank billets by some typical metrics of "jointness" (such as number and frequency of interactions with non–own-service organizations, percentage of time spent on various tasks, and importance of tasks to jobs). Because many different types of knowledge might be required for a joint duty assignment, we did not attempt to rank billets by the different types of knowledge required for or gained through an assignment.

However, we used the education and experience requirements and the types of experience afforded by the billets to rank the billets, by JDAL category and major billet organization (see Table 6.10). We use this information in Chapter Eight to provide an illustration of how such data can be used to determine demand for and supply of joint duty billets.

Not unexpectedly, JDAL billets rank first on every indicator. After all, the indicators reflect some of the criteria used to identify billets for the JDAL. Service-nominated billets rank behind other non-JDAL billets both in terms of gaining experience in certain areas and in the percentage of billets requiring joint education or experience. However, even among these service-nominated billets, 70 percent of them were identified as needing JPME II or

Figure 6.12
Percentage of Officers Who Agreed or Strongly Agreed that Their Position Provides Significant Experience in One, Two, or Three Key Areas, by Major Billet Organization

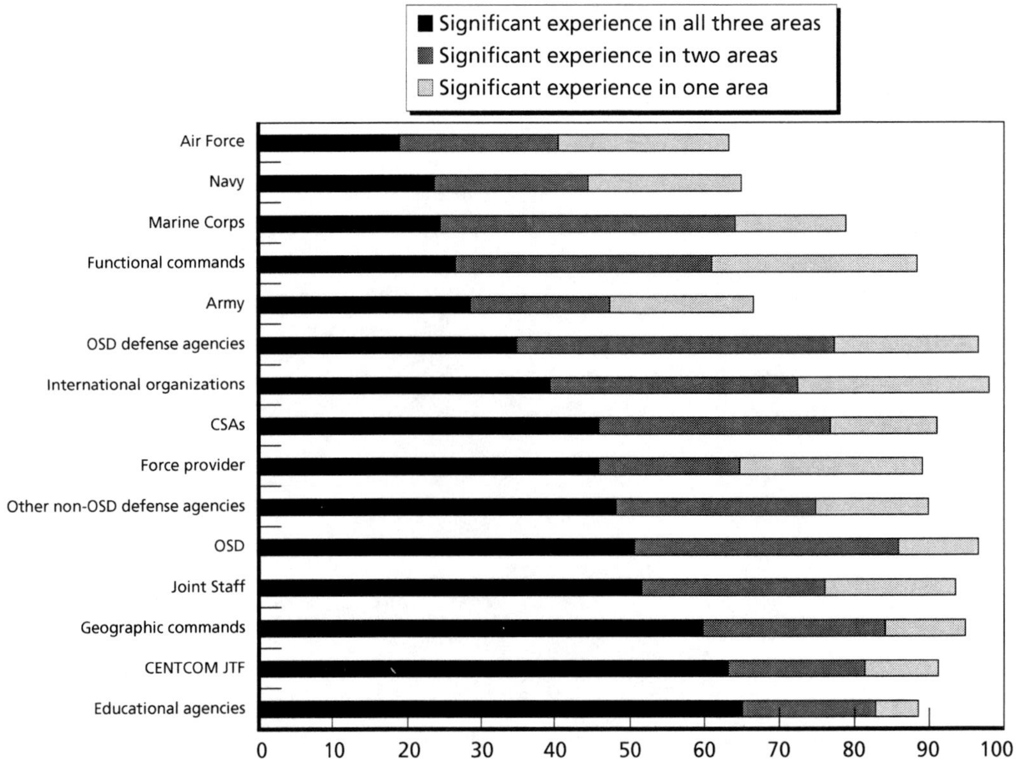

RAND *TR349-6.12*

prior joint experience. When we examine the rankings by billet organizations, the educational agencies, Joint Staff, OSD staff, and functional commands rank high on most indicators of experience gained in certain areas or in the need for joint experience and education. The rankings should be seen in light of the fact that well over 80 percent of all the non-service billets and Marine Corps billets were identified as requiring JPME II or prior joint experience. CENTCOM JTF billets and billets in the geographic commands rank second and third on providing significant experience in multiservice, multinational, and interagency matters.

Table 6.10
Rankings of Billet Categories Based on Selected Indicators of Education and Experience Required for and Experience Gained Through an Assignment, by JDAL Category and Major Billet Organization

	Rank Based on Percentage of Officers Reporting that JPME II Was Required or Desired for Assignment	Rank Based on Percentage of Officers Reporting that Prior Joint Experience Was Required or Desired for Assignment	Rank Based on Percentage of Officers Reporting Having Gained Significant Experience in Multiservice Matters	Rank Based on Percentage of Officers Reporting Having Gained Significant Experience in Multinational Matters	Rank Based on Percentage of Officers Reporting Having Gained Significant Experience in Interagency Matters	Rank Based on Percentage of Officers Reporting Having Gained Significant Experience in All Three Areas
JDAL Category						
JDAL billets	1	1	1	1	1	1
Non-JDAL billets in external organizations	2	2	2	2	2	2
Service-nominated billets	3	3	3	3	3	3
Major Billet Organization						
Joint Staff	5	5	2	7	5	4
OSD	4	3	1	5	1	5
CENTCOM JTF	10	7	7	2	8	2
Army	13	13	14	10	14	11
Navy	15	15	13	13	11	14
Air Force	14	14	15	14	12	15
Marine Corps	12	9	9	12	15	13
International organizations	9	11	12	1	13	9
CSAs	11	8	11	6	4	8
Other non-OSD defense agencies	8	10	8	8	2	6
OSD defense agencies	6	6	4	11	3	10
Educational agencies	1	1	3	4	7	1
Geographic commands	3	4	5	3	6	3
Force provider	2	2	6	9	9	7
Functional commands	7	12	10	15	10	12

NOTE: Organizations with equal percentages of respondents fitting the criteria received the same ranking.

CHAPTER SEVEN

Optimal Length of Joint Duty Assignments, Value of Temporary Assignments, and Other Issues Regarding Duties and Experience

This chapter presents respondents' answers to questions about the optimal length of joint duty assignments, the value of temporary JTF assignments, the ability of civilians or officers from another service to perform the tasks associated with a billet, and whether their experience differed from that of their predecessors and whether their successor's experience would likely be different from their own.

Optimal Length of Joint Duty Assignments

The length of duty tours and opinions about optimal lengths of duty tours differed by whether officers served in internal service billets or other billets. We use the median here to represent the typical response to avoid the effect of outliers.

Currently, most JDAs are 36 months in length. Based on the survey responses, the median planned length of assignment for most officers was 36 months for those assigned to JDAL billets and non-JDAL billets in external organizations and, on average, most respondents believed that this was the optimal length of time for these tours of duty and for permanent joint duty assignments. On average, officers in JDAL billets reported that it took about five months to become comfortable operating in a joint environment. That time was a little shorter for those in non-JDAL billets in external organizations—about three months. The 25th percentile was two months, and the 75th percentile was six months; therefore, the middle 50 percent of officers reported that it took between two to six months to become comfortable in a joint environment.

Those assigned to CENTCOM JTF billets reported becoming comfortable working in a joint environment within a very short period of time—just one month. The optimal length of assignment for respondents in these billets—12 months—was also shorter than the optimal length of time reported by those in other billets.

There was no difference in the responses by whether individuals had received credit for JPME I or JPME II. Higher-ranked officers appeared to become comfortable in billet assignments more quickly than other officers.

Among service-nominated billets, the tour of duty for officers was shorter—24 months—and, like the others, officers in these billets believed that this period of time was the optimal length for the tour of duty assignment and for permanent joint duty assignments. The median length of time reported by these officers for becoming comfortable in the joint environment was four months—somewhat shorter than the six months reported by officers in non-service billets.

Officers serving in assignments that were one to two years in length reported that the optimal length of a permanent JDA was about 24 months, while those serving in longer assignments (the majority of whom were serving three-year tours) believed that the optimal length of permanent assignments was 36 months.

All officers held similar views about the optimal length of a temporary joint task force assignment. Regardless of where they were assigned, on average, officers believed that the optimal median length of such tours should be six to seven months, and the length of time cited by the majority of respondents ranged from six months to more than one year. About 25 percent of officers reported that that the optimal assignment length was about six months, while another 25 percent believed that the assignment should last one year or longer.[1]

Value of Temporary JTF Assignments

Respondents who had served at least two months in a permanent joint duty assignment were also asked whether they had served in a temporary JTF assignment since September 2001. About 20 percent of those in JDAL billets and non-JDAL billets in external organizations and about 14 percent of those in service-nominated billets met these criteria. These respondents were asked to rate the value of these JTF assignments in obtaining an understanding of the joint environment, relative to the permanent joint duty assignment billet. Overall, about 23 percent reported that the JTF assignment allowed them to become familiar with the joint environment considerably more quickly, another 23 percent reported that that the JTF assignment allowed them to become familiar with the joint environment slightly more quickly, and 40 percent believed that the two experiences were about the same. About 14 percent reported the converse to be true; they believed that the permanent duty assignment was more valuable in providing a quicker understanding of the joint environment.

Ability of a Civilian or an Officer from Another Service to Perform Respondents' Duties

When asked whether a civilian could perform their duties just as effectively, about 30 percent of respondents overall agreed or strongly agreed that the civilian could. When asked whether their job required unique knowledge of their own service and, as such, could not be performed by an officer from another service, close to 50 percent disagreed or strongly disagreed that an officer from another service could not do the job. We find some interesting and not unexpected differences given the billet organization to which officers were assigned. Figure 7.1 shows the percentage of officers who agreed or strongly agreed that a civilian could perform their duties and the percentage of those who disagreed or strongly disagreed that an officer from another service could not perform their duties; for ease of exposition in the figure, we label the latter "officer from another service could perform duties."

---

[1] Responses varied among those serving in internal service billets. The range between the 25th and the 75th percentiles (the middle half of the distribution) was six to 12 months for Army and Marine Corps officers, four to 8.5 months for Navy officers, and four to six months for Air Force officers.

Figure 7.1
Percentage of Officers Who Agreed or Strongly Agreed that Their Duties Could Be Performed by a Civilian or by an Officer from Another Service, by Major Billet Organization

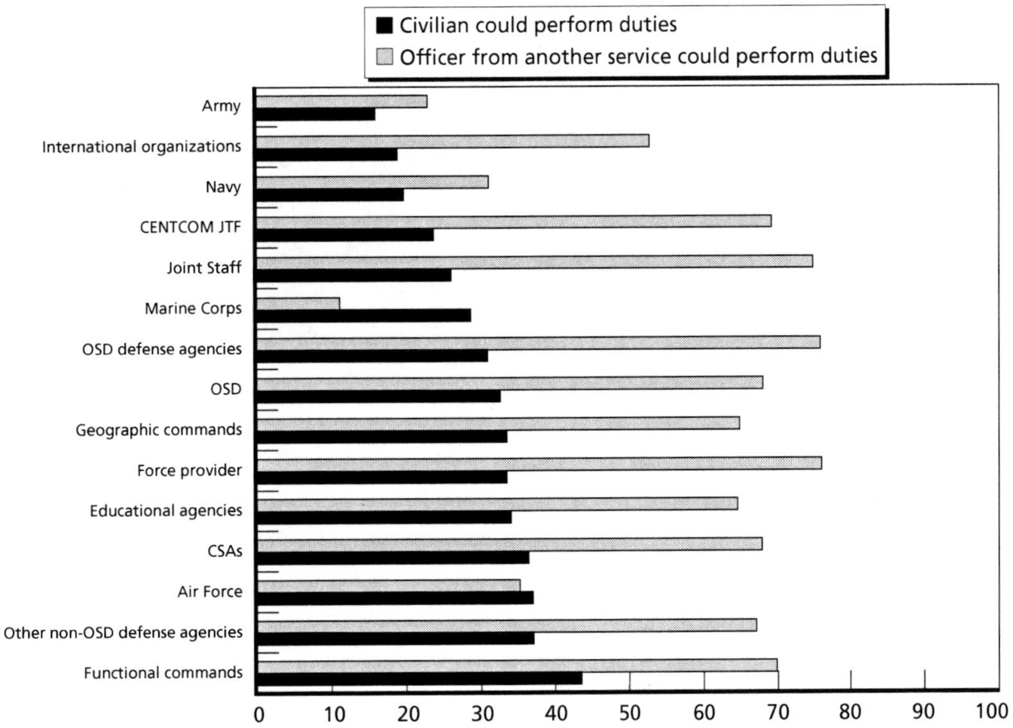

RAND *TR349-7.1*

Officers in the Army, Navy, international organization, and CENTCOM JTF billets were the least likely to believe that their duties could be performed by a civilian, while those in the functional command, Air Force, and non-OSD defense agency billets were the most likely to believe that they could. Officers in internal service billets believed that their positions required knowledge of their own service and, as such, could not be performed by an officer from another service. Only 11 percent of those in Marine Corps billets and 23 percent of those in Army billets agreed that an officer from another service could perform their duties, and this outcome is consistent with what we had seen in Chapter Six about officers needing to draw on their knowledge of their own service's capabilities to carry out their jobs. Seventy percent or more of officers assigned to the functional command, Joint Staff, OSD defense agency, and force provider billets reported that their duties could be carried out by an officer from another service.

Only about 8 percent of flag officers and 21 percent of O-6s agreed that a civilian could perform their duties just as effectively as they can. Flag officers also were less likely than mid-grade officers (27 percent compared with 50 percent) to report that an officer from another service could carry out their assignments.

Uniqueness of Experience

As a lead-in to questions about the skills and competencies that are important in their assignments, for both themselves and for successors, officers were asked whether their experience differed from that of their predecessors as a result of current events and whether their successors' experience was likely to be different from their own experience.

There was not much difference by JDAL category overall, with just about half reporting that their experience differed from that of their predecessors and between 22 and 25 percent reporting that their successors' experience was likely to be different from their own experience. However, there are interesting differences in the responses depending on where the officers are assigned, as shown in Figure 7.2, which is arrayed by the percentage of officers reporting that their experience was different from that of their predecessors. Officers assigned to international organizations were the most likely to report that their experience differed from that of their predecessors. This is not surprising, given the enormous changes in the international environment over the past five years. Navy officers were the least likely to report as such. Officers serving in CENTCOM JTF billets were the most likely to report that their position would be different for their successors because their own was based on unusual current events, while officers assigned to educational agencies were the least likely to report that their positions would be different. CENTCOM JTF officers work on joint task forces set up to address specific issues, so it is not surprising they believe that their successors' experience will differ from theirs.

Figure 7.2
Percentage of Officers Reporting that Their Experience Differed from That of Their Predecessors and that Their Successors' Experience Will Likely Differ from Their Own, by Major Billet Organization

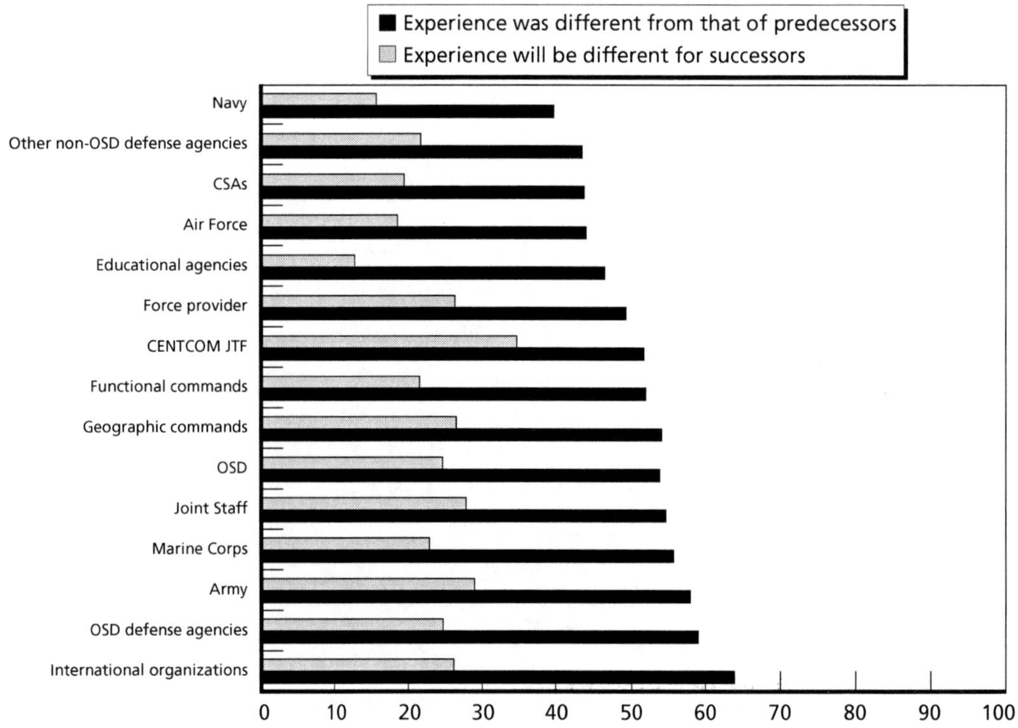

RAND *TR349-7.2*

Summary

Table 7.1 ranks the billets using the indicators of the ability of civilians and the ability of officers from another service to carry out the duties of the billets. Not surprisingly, non-JDAL billets in external organizations rank first on both indicators.

CSA, other non-OSD defense agencies, force provider, and functional command billets ranked high on these indicators. Somewhat unexpectedly, officers in Air Force billets ranked high in their belief that civilians would be able to substitute for them in their positions. Officers at the Joint Staff and OSD defense agencies tended to believe that officers from another service could substitute for them.

Table 7.1
Billet Categories Ranked According to Selected Indicators Relating to Substitutability of Civilians and Other Service Officers, by JDAL Category and Major Billet Organization

	Rank Based on Percentage of Officers Who Agreed/Strongly Agreed that a Civilian Could Perform Their Duties	Rank Based on Percentage of Officers Who Agreed/Strongly Agreed that an Officer from Another Service Could Perform Their Duties
JDAL Category		
JDAL billets	2	2
Non-JDAL billets in external organizations	1	1
Service-nominated billets	3	3
Major Billet Organization		
Joint Staff	11	3
OSD	8	6
CENTCOM JTF	12	5
Army	15	14
Navy	13	13
Air Force	3	12
Marine Corps	10	15
International organizations	14	11
CSAs	4	7
Other non-OSD defense agencies	2	8
OSD defense agencies	9	2
Educational agencies	5	10
Geographic commands	7	9
Force provider	6	1
Functional commands	1	4

NOTE: Organizations with equal percentages of officers fitting the criteria received the same ranking.

CHAPTER EIGHT

Using Survey Data to Estimate Current and Projected Demand for and Supply of Joint Officers

This report provides an overview of the results of the 2005 Joint Officer Management Census survey. The survey was designed to elicit information on joint billets on the JDAL, potential joint billets in external organizations, and internal service-nominated billets that require or provide joint experience. As stated earlier, we examined the responses of more than 21,000 officers regarding their assignments, including the extent to which officers believe that their assignments provide them with joint experience or require them to have had prior joint education, training, or experience and whether and how these answers differ across organizations and/or services where the billets are located.

We used the survey data to develop rough estimates of the demand for and supply of joint officers now and in the future. We caution that these estimates are based on simplistic assumptions, and considerable work remains to be done to refine the analyses shown here. Therefore, these findings should be regarded as merely suggestive. Nonetheless, the findings are useful in illustrating how the data presented in this report can be combined with other data to support more-analytical models of joint officer supply and demand.

Current and Projected Joint Officer Supply

As discussed in Chapter One, the Goldwater-Nichols Act requires the Secretary of Defense to develop a definition of a JDA and to publish a JDAL. The JDAL includes those positions at organizations outside the individual services that provide significant experience in joint matters. Joint duty consideration is limited by policy to billets in pay grades of O-4 or higher. In some organizations (OSD, the Joint Staff, and the unified commands), 100 percent of such positions were placed on the JDAL, whereas 50 percent of the positions in each defense agency were placed on the list. The law specifically prohibits positions in service organizations from receiving joint duty credit. These parameters led to a list of approximately 9,000 positions designated as JDAs. Officers receive joint credit by serving in one of these billets for a specified period of time. The current system is time-based and billet-based to provide a means for determining officer qualifications.

Goldwater-Nichols created a new category of officers, called JSOs, who are particularly trained in joint matters. Prerequisites to becoming a JSO include successfully completing a program at a JPME school and, after completing such a program, serving a full tour in a JDA.

Thus, the current and projected supply of joint officers depends crucially on the (1) availability of the resources that create joint-qualified officers—JPME II seats and JDAs and (2) retention of such officers.

Current Supply of Joint Officers[1]

Officers with Joint Experience. The percentage of officers (O-4 and above) with some previous joint experience has been increasing over time. Excluding health occupations, as of 2002, between 30 and 40 percent of these officers in each service had been assigned to a JDA (including those with partial credit or "currently serving" in a JDA at that time). The percentage of O-4s who have ever served in a JDA holds reasonably steady, at between 15 and 25 percent in each service, but the percentages for O-5 and O-6 are much higher. Jointness accretes over time as officers continue in service and are promoted. In 2002, the percentage of non–health-care officers in grades O-5 and O-6 with a joint assignment was 55 percent in the Navy, more than 60 percent in the Army and Air Force, and more than 70 percent in the Marine Corps.

Looking at the O-5 and O-6 combined officer population by occupation yields further understanding of which officers tend to have had joint experience. For each service, intelligence officers are considerably more likely than officers in other billets to have experienced a joint assignment (70–80 percent in 2002). The percentage of tactical operations officers (in all services) and engineering/maintenance officers (in all but the Navy) with joint experience have increased sharply over time, and stood at between 50 and 60 percent in 2001. Data show that for three of the four services, 65–75 percent of the tactical operations O-6s had been exposed to joint experience, although only 40 percent had been accredited with jointness. The accretion of joint experience in the services is also partly determined by the leaving rates of these officers. In 2002, between 5 and 15 percent of O-5s and O-6s who had some joint experience, who had served in a JDA, or who were JSOs left the service.

Thus, the current availability of educated officers with joint experience is constrained in the steady state by resource availability—JPME II seats and JDAs—and by officers' behavior.

Billets Available to Provide Joint Experience. The current number of billets on the JDAL is around 9,000; they are the only JDAs for which an officer can get joint duty credit. However, in the present environment, more officers are gaining credit for joint experience through cumulative service in designated JTFs and through use of SecDef waiver authority. But this "spike" in experience will dissipate over time as officers leave the service. Ultimately, the number of joint-experienced officers provided in a billet-based system depends on the number of qualifying billets and the length of time served in them.

JPME Seats to Provide Joint Education Credit. As of this writing, there are 1,172 seats currently authorized and certified for JPME II credit. They include 124 seats at the National War College (NWC), 180 seats at the Industrial College of the Armed Forces (ICAF), 840 seats at the Joint and Combined Warfighting School (JCWS) and 28 seats at the Joint Advanced Warfighting School (JAWS), both at Joint Forces Staff College (JFSC).

Thus, the maximum number of JPME graduates that could be assigned to JDAs each year is 1,172. However, given the need for some graduates within the services, it is unreasonable to expect that all of these graduates will be assigned to JDAs. An earlier RAND study (Thie et al., 2005) concluded that, realistically, about 83 percent of JPME II graduates could be assigned to JDAs.

[1] This section is largely based on Thie et al., 2005.

Assessment. Using a number of assumptions,[2] the Thie et al. study showed that there is a relationship between the ability to fill joint positions and the number of JPME graduates, the assignment policy for JPME graduates, the number of JSOs assigned, and other factors. Thus, under the assumption that 1,000 JSOs serve in critical or non-critical JDAs, the maximum joint need that could be met is approximately 10,400 positions. Also, there has been a leveling off of joint experience in the services. This is not surprising, given that the seats for JPME II are limited and the opportunity to have a joint duty assignment is also limited by the number of billets on the JDAL.

Projected Supply of Joint Officers

JPME Seats. The projected supply of joint officers will increase at least partly because of the increase in the number of eligible JPME II seats. The number of JPME II seats will increase to 1,908 as senior-level service programs become eligible for future PAJE accreditation for JPME II. These seats include 227 seats at the U.S. Army War College; 312 seats at the U.S. Navy College of Naval Warfare; 13 seats at the U.S. Marine Corps War College; and 184 seats at the U.S. Air Force Air War College.

While there will be more JPME II graduates available in the future, more of them will be from colleges that send their graduates as much to service as to joint positions. As a result, the likely assignment outcome from all schools will be around 60 percent of graduates. Using the same methodology as that described above, a joint need of about 11,830 positions could likely be met. More billets could be supported by further increasing the number of JPME II seats, by increasing the percentage of JPME grads assigned directly to joint positions, or by increasing the number of JSOs assigned to joint positions. For example, more JPME II graduates who serve in qualifying billets produce more JSOs. Using the existing estimating relationship, another 267 joint billets could be filled for each 100 JSOs produced and then later assigned to another joint assignment (e.g., 1,100 versus 1,000). As part of their ongoing look at demand for and supply of joint qualified officers, RAND is assessing these relationships and how these new patterns of use would fit into career paths.

Billets Available to Provide Joint Experience. Current law restricts in-service billets from receiving joint credit. One frequent criticism of the system is that officers are serving in other assignments that provide a rich joint experience but do not grant the officer joint credit. Likewise, there are officers serving in assignments on the JDAL that may not provide what some would consider a joint experience, either because of the content or context of their work or because of limited interaction with other services, nations, or agencies.

The 2005 JOM Census Survey provides some interesting data on the kinds of experiences that non-JDAL billets provide. As we showed in Chapter Six, respondents were asked whether their billet provided "significant experience" in multiservice, multinational, and/or interagency matters. Of the more than 20,000 respondents who answered this set of questions, approximately 16,000 (79 percent) agreed or strongly agreed that their billet provided them with one or more of these kinds of experiences. Among these 15,947 billets, a little less than half—7,528—provided experience of all three kinds (multiservice, multinational, and interagency); 4,776 provided two kinds of experience; and 3,623 provided experience in at

[2] These assumptions include: JSOs serve in either critical or noncritical positions; average JDA tour length is three years; 12.5 percent of the JDAL is filled by COS Exceptions; and no COS Exceptions return to JPME.

least one area. Table 8.1 shows the number of billets providing significant multiservice, multinational, or interagency experience, by JDAL category.

Whether billets provide significant experience in multiservice, multinational, and/or interagency matter is obviously only one of several criteria that could be used to determine a valid joint experience. The subsequent analysis will use billet-level data to assess the degree of jointness the billet provides to officers, using multiple criteria such as amount or level of interaction and duties, types of experience, knowledge, and education required for the job, and other factors.

Nonetheless, the implication for projected joint officer supply is significant if a new JDAL or other means to determine joint qualifications is based on a more liberal interpretation of the current philosophy and includes either all officers at joint organizations, officers in selected internal service billets, or both.

Tour Lengths. The current JDA tour length is 36 months, and most officers reported that this was the optimal tour length. If the tour length is decreased, then the projected supply of joint officers would increase as officers cycle through JDAs faster than before. Future RAND analysis will consider the implications of shorter tours of duty on projected supply.

Assessment. The projected supply of joint officers should increase, given the increase in JPME II seats. If a more liberal interpretation of joint duty assignments is adopted or shorter tour lengths become the norm, then the services should experience a marked increase in "jointness" of the officer corps.

Demand for Joint Officers

A strategic workforce plan (1) determines levels of critical characteristics needed to carry out missions and goals, (2) assesses current and future availability, and (3) develops policies to eliminate gaps between demand and supply. It is important to understand future need for joint officer resources to be able to produce the resources to match the determined need. We can infer need for one or the other of two critical workforce characteristics—joint experience and joint education—from three requirements in the law: First, there is the requirement for officers to have completed a JDA prior to promotion to general or flag officer, which sets a

Table 8.1
Number of Billets Providing Significant Experience in Multiservice, Multinational, or Interagency Matters, by JDAL Category

JDAL Categories	Number of Billets Providing at Least One Kind of "Joint" Experience	Total Number of Non-Missing Cases in Survey
JDAL billets	5,589	5,937
Non-JDAL billets in external organizations	3,833	4,271
Service-nominated billets		
Army	2,467	3,708
Navy	1,274	1,962
Air Force	2,736	4,317
Marine Corps	48	61
Total service-nominated billets	6,525	10,048

requirement for joint experience for most of the approximately 900 general and flag officer positions.[3] Second, there is a requirement to fill 800 critical positions with JSOs, which means that officers in those positions need to have successfully completed JPME II and a prior JDAL assignment. Third, there is the requirement to fill at least half the JDAL positions with a JSO or JSO nominee; thus, at least 50 percent of JDAL positions need to be filled with officers who have completed JPME II. Currently, this need for joint-officer resources is identified only by the 800 critical billets; however, data gathered in the 2005 JOM survey suggests that the need for such resources is more widespread.

In the 2005 JOM survey, as we showed in Chapter Six, respondents identified, among other aspects, the need for joint education and prior joint experience for successfully carrying out the duties of the billet.

Tables 8.2 and 8.3 display the survey responses for JPME II and prior joint experience by JDAL category for officers in billets at the O-3–O-6 ranks. In Table 8.3, we combine data for JPME II and prior joint experience—specifically, we combine the data for those who answered "required" for both questions and those who answered "not helpful" for both questions—and we report variations of "required" and "desired" responses.

Need for JPME II

Among the approximately 20,000 respondents in billet grades of O-3–O-6, a large number (about 9,500) had not had JPME II and, therefore, did not answer the question. Among those who did answer the question,

- approximately 2,200 believed that JPME II was "required" for their billet, and the majority (1,386) were in JDAL billets
- close to 6,000 believed that JPME II was "desired," about 1,300 of whom were in non-JDAL billets in external organizations and about 2,100 of whom were in service-nominated billets
- approximately 1,700 believed that JPME II was "not helpful."

Need for Prior Joint Experience

Large numbers of respondents believed that prior joint experience was required or desired to carry out their assignments successfully.

- Approximately 2,500 said prior joint experience was "required" for their billet, about 1,000 of whom were in JDAL billets.
- Approximately 12,500 responded that such experience was "desired," about 3,000 of whom were in non-JDAL billets in external organizations and about 5,400 of whom were in service-nominated billets,
- Approximately 4,100 said such experience was "not helpful," and most, not surprisingly, were in service-nominated billets.

---

[3] Some officers, such as doctors, are exempt from this requirement.

Table 8.2
Number of Billets with Need for JPME II or Prior Joint Experience, by JDAL Category, O-3–O-6

	JPME II					Prior Joint Experience			
	Required	Desired	Not Helpful	Not Sure—Have No JPME II Experience	Missing	Required	Desired	Not Helpful	Missing
JDAL billets	1,386	2,365	337	1,839	201	1,030	4,236	660	202
Non-JDAL billets in external organizations	401	1,246	276	2,154	206	559	2,882	636	206
Service-nominated billets									
Army	193	809	312	2,099	232	385	2,083	945	232
Navy	47	332	284	1,027	72	147	898	645	72
Air Force	144	950	484	2,303	172	335	2,351	1,195	172
Marine Corps	1	25	5	18	7	7	34	8	7
Total service-nominated billets	385	2,116	1,085	5,447	483	874	5,366	2,793	483
Total	2,172	5,727	1,698	9,440	890	2,463	12,484	4,089	891

Table 8.3
Number of Billets with Combined Need for JPME II and Prior Joint Experience, by JDAL Category, O-3–O-6

	Need for JPME II/Prior Joint Experience		
	Number of Officers Responding that Both JPME II and Prior Joint Experience Are Required for the Billet	Number of Officers Responding that JPME II and Prior Joint Experience Are Either Required or Desired or that Both Are Desired for the Billet	Number of Officers Responding that Neither JPME II nor Prior Joint Experience Is Helpful for the Billet
JDAL billets	470	3,035	109
Non-JDAL billets in external organizations	167	1,398	98
Service-nominated billets			
Army	127	825	157
Navy	27	342	172
Air Force	69	959	310
Marine Corps	0	24	3
Total service-nominated billets	223	2,150	642
Total	860	6,583	849

NOTE: This table excludes respondents with missing, "unsure," or "not helpful" responses on either question, unless both responses were in the "not helpful" category.

Need for JSOs

If we examine the intersection between the responses to the two questions (whether prior joint experience was required and whether it was desired), we get an estimate of the numbers of billets for which both joint education and prior joint experience is believed to be essential, indicating a need for JSOs.

- Approximately 860 billets were identified as needing *both* joint education and prior joint experience, of which only about half were JDAL billets. This indicates that there are non-JDAL billets both in the external organizations and in the services that may need JSOs.
- More than 6,500 billets were identified as billets for which such education and prior joint experience may be desired, or at least one of them required and the other desired, for successfully carrying out the assignment.
- Only about 849 of all respondents reported that neither education nor prior joint experience was helpful for their billet, approximately 100 of whom were in JDAL and non-JDAL billets in external organizations and about 600 of whom were in the service-nominated billets.

Estimating Future Needs

These data can be used to make rudimentary estimates of the number of billets to be filled by officers with joint experience, with joint education, or both (JSOs). These estimates will be refined as part of the continuing analysis of the 2005 JOM Survey.

The most conservative assumption is that, because the data collection was designed to be a census of incumbents and supervisors, those who responded accurately described billet needs, and those billets for which there is no response do not need officers with education and experience characteristics to fill them. In this case,

- 860 billets must be filled by JSOs
- about 7,040 positions (either required or desired) should be filled by officers with JPME II (not including JSOs)
- about 14,100 positions (either required or desired) should be filled by officers with joint experience (not including JSOs).

A more likely assumption is that respondents on the JDAL have similar billet characteristics to those who did not respond. In that case, we project billet needs to the full current JDAL (but not to the other organizational clusters) with the following results:

- A total of 1,040 billets must be filled by JSOs.
- Approximately 8,300 positions (either required or desired) should be filled by officers with JPME II (not including JSOs).
- Approximately 15,900 positions (either required or desired) should be filled by officers with joint experience (not including JSOs).

A more extreme assumption is that respondents not now on the JDAL in the two other organizational clusters have similar billet characteristics to those who did not respond. In that case, we project billet needs for all organizational clusters, with the following results:

- A total of 1,210 billets must be filled by JSOs.
- Approximately 9,900 positions (either required or desired) should be filled by officers with JPME II (not including JSOs).
- Approximately 19,900 positions (either required or desired) should be filled by officers with joint experience (not including JSOs).

These data provide a first cut at quantifying the need for JSOs, joint education, and experience across this set of billets. It is clear that non-JDAL billets and service-nominated billets would benefit from such prior joint experience and education.

- Approximately 860 to 1,210 billets would benefit from both education and joint experience.
- Approximately 7,040 to 9,900 billets would benefit from JPME II.
- Approximately 14,100 to 19,900 billets would benefit from prior joint experience.

These assumptions have used the total number of JDAL billets identified in our survey—about 8,500. Given that the actual number of JDAL billets is higher—in excess of 9,700—the total number of estimated billets requiring JSOs, JPME II, and/or prior joint experience is likely to be higher than that estimated here. We chose the more conservative numbers simply because they based directly on data from the surveyed billets.

This chapter presents what is at best a cursory analysis based on raw numbers. We have not completely accounted for nonresponse and the fact that the survey excluded some organizations or failed to fully account for all billets that it was meant to include because of an outdated sampling frame. A billet-by-billet analysis is needed (and is under way) to understand and measure more accurately the need for JSOs, prior joint experience, and education. Nonetheless, these raw data indicate a substantial need for experience and education in billets that are not on the JDAL. For example, the need for joint experienced and educated officers at the grade of O-6 was seen in service billets for commanders, center directors, plans and policy officers, and educators.

The 2005 JOM Census Survey was designed as a one-time data collection effort. However, Thie et al. (2005), the precursor work to this study, emphasized that a long-term strategic approach would require the services and the Joint Staff to standardize and to routinely collect and maintain information about billet requirements. This requirement could become part of a DoD Directive or DoD Instruction. The analysis in this chapter shows how the JOM survey data can be used to inform ongoing joint officer management and development issues.

2005 Joint Officer Management Census Survey Form

REPORT CONTROL SYMBOL (RCS): DD-P&R(OT)2196

JOINT OFFICER MANAGEMENT (JOM) CENSUS

PRIVACY ACT STATEMENT

AUTHORITY. 10 U.S.C. 136; 10 U.S.C. 1782; 10 U.S.C. 2358; 10 USC 531.

PRINCIPAL PURPOSE(S). TO CAPTURE DATA REQUIRED FOR A BILLET-LEVEL ANALYSIS OF JOINT CHARACTERISTICS. THE INFORMATION OBTAINED IN THIS CENSUS WILL BE USED TO ASSIST THE OFFICE OF THE SECRETARY OF DEFENSE IN DEVELOPING A STRATEGIC PLAN FOR JOINT OFFICER MANAGEMENT AND JOINT PROFESSIONAL MILITARY EDUCATION THAT LINKS JOINT OFFICER DEVELOPMENT AND ACCOMPLISHMENT TO THE OVERALL MISSION AND GOALS OF THE DEPARTMENT OF DEFENSE.

ROUTINE USES. NONE.

DISCLAIMER. THIS CENSUS HAS BEEN APPROVED FOR INFORMATION COLLECTION AND HAS BEEN ASSIGNED A REPORT CONTROL SYMBOL (RCS, DD-P&R(OT)2196. IN ACCORDANCE WITH DoD 8910.1-M, "DoD PPROCEDURES FOR MANAGEMENT OF INFORMATION REQUIREMENTS." THE EXPIRATION DATE FOR RCS DD-P&R(OT)2196 IS APRIL 30, 2008. THE APPROVED SD FORM 455, "REQUEST FOR APPROVAL OF INFORMATION COLLECTION" HAS BEEN PLACED IN THE OFFICIAL FILE.

DISCLOSURE. VOLUNTARY. HOWEVER, FAILURE TO PROVIDE THE INFORMATION MAY RESULT IN YOUR VIEWS NOT BEING REPRESENTED IN THE FINAL RESULTS AND OUTCOMES.

Office of the Secretary of Defense

Enter your ID and Password forwarded to you by your organization's point of contact. Your census key is **not** your ID or your Password; instead it will be entered later in order to identify your particular billet.

ID: [ ]

Password: [ ]

[Enter Census]

For login-census key tips, click --> here

This census is best viewed using Internet Explorer.

Office of the Secretary of Defense

Welcome Statement from the Honorable David S. C. Chu

There have been great strides in the joint arena since the enactment of the Goldwater-Nichols Act of 1986; but we must improve how we manage the joint officer force. Congress, the General Accountability Office and our own independent reviews all argue for this improvement.

In partnership with the Joint Staff/J-1, my office is undertaking a "clean slate" strategic review of joint officer management. It includes an assessment of the characteristics of joint positions; a comprehensive exploration of the full set of needs for joint experience; and how we refresh and expand our understanding of jobs that provide joint experience. The census you are about to take is a cornerstone for the review.

I ask you to take the next 15 minutes to focus on this census. I also ask that you encourage those in your organization who are asked to complete the census to do the same. In order to achieve meaningful results, we need close to 100 percent return. The time you invest will provide input and direction to significant legislative and policy decisions in the near future.

Thank you for your help & and your service.

David S. C. Chu
Under Secretary of Defense
(Personnel and Readiness)

Next ▸

Q-1 Please enter census key below:

Census Key: [ ]

◂ Previous Next ▸

For login-census key tips, click here

Confirm billet

Q-2 Is this the billet you have been asked to evaluate?

Billet Key	: ZZZZZZZZ
Organization	: Navy
Organization Sub-Element	: UNKNOWN
Unit Identification	: 00019
Unit Identification Clear Text	: COMNAVAIRSYSCOM
Organization Position Identifier	: 0002305
Other Billet Specific Information Text	: ZZZZZZZ
Major Command	: UNKNOWN
Work Center Name	: ZZZZZZZZ
Work Center Identifier	: ZZZZZZZZZZ
Work Center Position Name	: Commander, Naval Air Systems Command
Work Center Position Identifier	: 00300
Position GEOLOC Code	: ZZZZ
Position GEOLOC	: UNKNOWN

Note: some data for your billet may be unavailable and, therefore, may be represented by "ZZZZZZZZZ", "NO DATA FOUND" or a blank return. Please disregard these entries and review the data that is available to make an assessment that you were given the correct census key for your billet.

○ Yes
○ No

◂ Previous Next ▸

For troubleshooting tips, click -->here
To email technical support, click here: jomhelp@lmi.org

Q-3 Are you the:

*Select **one** answer*

- (●) Person occupying the billet? Please ensure that you answer all of the following questions based on the confirmed billet, to include any TAD/TDY experiences during your service in the billet. If your billet is a temporary joint task force (JTF) billet, answer all questions based solely on your JTF experience unless instructed otherwise.
- ( ) Supervisor of the billet?
- ( ) Another person designated to complete the survey?

◂Previous Next▸

For troubleshooting tips, click -->here
To email technical support, click here: jomhelp@lmi.org

Q-4 How many **MONTHS** have you been assigned to this billet?

*Please enter a **whole** number*

[ ]

◂Previous Next▸

For troubleshooting tips, click -->here
To email technical support, click here: jomhelp@lmi.org

Q-5 Is the billet you have been asked to evaluate a temporary JTF billet?

*Select **one** answer*

- ( ) Yes
- ( ) No

◂Previous Next▸

For troubleshooting tips, click -->here
To email technical support, click here: jomhelp@lmi.org

Q-6 Please enter the total number of **WEEKS** you have spent on TDY / TAD (NOT including training and education) during this assignment.

*Please enter a **whole** number*

[0]

◂Previous Next▸

For troubleshooting tips, click -->here
To email technical support, click here: jomhelp@lmi.org

Q-7 Please enter the total number of **WEEKS** you have spent on TDY / TAD due to training or education during this assignment:

*Please enter a **whole** number*

[0]

◂Previous Next▸

For troubleshooting tips, click -->here
To email technical support, click here: jomhelp@lmi.org

Questions 9 through 11 are optional. Answers will not be used for the JOM study.

Q-8 What is your pay grade?

*Select **one** answer*

Please Select

Q-9 What is your gender?

*Select **one** answer*

- ○ Male
- ○ Female

Q-10 Are you Spanish/Hispanic/Latino?

*Select **one** answer*

- ○ No, not Spanish/Hispanic/Latino
- ○ Yes, Mexican, Mexican-American, Chicano, Puerto Rican, Cuban, or other Spanish/Hispanic/Latino

Q-11 What is your race?

Mark one or more races to indicate what you consider yourself to be.

- ☐ White
- ☐ Black or African-American
- ☐ Native American or Alaska Native
- ☐ Asian (for example: Asian Indian, Chinese, Filipino, Japanese, Korean, or Vietnamese)
- ☐ Native Hawaiian or other Pacific Islander (for example: Samoan, Guamanian, or Chamorro)
- ☐ Other

Q-12 How many **YEARS** have you been a commissioned officer?

*Please enter a **whole** number.*

Q-13 Are you:

*Select **one** answer*

- ○ Active Duty List Officer
- ○ Reserve Component Officer

◂Previous Next▸

For troubleshooting tips, click -->here
To email technical support, click here: jomhelp@lmi.org

Q-14 Of the following list, what intermediate or senior level schools have you completed? Next to each school that you have completed, note your paygrade at time of completion.

	SELECT ONLY IF COMPLETED
JOINT AND COMBINED SCHOOLS	
National War College	-- Please Select --
Industrial College of the Armed Forces	-- Please Select --
School of Information Warfare and Strategy	-- Please Select --
Joint Forces Staff College	-- Please Select --
Joint and Combined Staff Officer School	-- Please Select --
Joint and Combined Warfighting School	-- Please Select --
US ARMY SERVICE SCHOOLS	
US Army War College (resident)	-- Please Select --
US Army War College (non-resident)/Dept. of Distance Education	-- Please Select --
Army Command and General Staff College (resident)	-- Please Select --
Army Command and General Staff College (non-resident)	-- Please Select --
US NAVY SERVICE SCHOOLS	
College of Naval Warfare	-- Please Select --
College of Naval Command and Staff (resident)	-- Please Select --
College of Distance Education (Navy intermediate-level college non-resident)	-- Please Select --
US AIR FORCE SERVICE SCHOOLS	
Air War College (resident)	-- Please Select --
Air War College (non-resident)	-- Please Select --
Air Command and Staff College (resident)	-- Please Select --
Air Command and Staff College (non-resident)	-- Please Select --
US MARINE CORPS SERVICE SCHOOLS	
Marine Corps War College	-- Please Select --
Marine Corps Command and Staff College	-- Please Select --
Marine Corps College of Continuing Education (Marine Corps Command and Staff College non-resident)	-- Please Select --

Q-15 If you have received credit for post-graduate education at another civilian or military institution not listed above (to include international institutions).

Please enter the name of the institution and the pay grade you held when you graduated in the boxes below.

INSTITUTION	Pay Grade
	-- Please Select --
	-- Please Select --
	-- Please Select --
	-- Please Select --

◂Previous Next▸

For troubleshooting tips, click -->here

To email technical support, click here: jomhelp@lmi.org

Q-16 Have you received credit for JPME I?

*Select **one** answer*

- ○ Yes
- ○ No
- ○ Not Sure

Q-17 Have you received credit for JPME II?

*Select **one** answer*

- ○ Yes
- ○ No
- ○ Not Sure

Q-18 What is your service and component?

*Select **one** answer*

- ○ U.S. Army
- ○ U.S. Army National Guard
- ○ U.S. Army Reserve
- ○ U.S. Navy
- ○ U.S. Naval Reserve
- ○ U.S. Air Force
- ○ U.S. Air National Guard
- ○ U.S. Air Force Reserve
- ○ U.S. Marine Corps
- ○ U.S. Marine Corps Reserve

◂Previous Next▸

For troubleshooting tips, click -->here
To email technical support, click here: jomhelp@lmi.org

Microsoft Internet Explorer

You are approximately one-quarter complete, by time required, for this census.

OK

Q-19 Select your primary AFSC, e.g., 11MX (qualified Mobility Pilot) from the following list.

*Select **one** answer*

-- Please Select --

◂Previous Next▸

For troubleshooting tips, click -->here
To email technical support, click here: jomhelp@lmi.org

Q-20 Select your primary Area of Concentration (AOC) code, e.g., 11A (Infantry) from the following list.

-- Please Select--

◂Previous Next▸

For troubleshooting tips, click -->here
To email technical support, click here: jomhelp@lmi.org

Q-21 Select your 4 digit officer designator code, e.g., 111X (Surface Warfare Officer) from the following list.

--Please Select --

◂Previous Next▸

For troubleshooting tips, click -->here
To email technical support, click here: jomhelp@lmi.org

Q-22	Select your 4 digit Military Occupational Specialty (MOS) code, e.g., 0802 (Field Artillery Officer).
	-- Please Select --

◂Previous Next▸

For troubleshooting tips, click -->here
To email technical support, click here: jomhelp@lmi.org

Q-23	Are you a Joint Specialty Officer (JSO)?
	○ Yes ○ No ○ Unsure

◂Previous Next▸

For troubleshooting tips, click -->here
To email technical support, click here: jomhelp@lmi.org

Q-24	Who is your 1st level supervisor (e.g., Rater or Reporting Senior)?	-- Please Select --
Q-25	Who is your 2nd level supervisor (e.g., Intermediate Rater, Additional Rater or Reviewing Senior)?	-- Please Select --
Q-26	Who is your 3rd level supervisor (e.g., Senior Rater or Reviewer)?	-- Please Select --
Q-27	In your current position, do you serve full time with members from another Military Department (i.e., Army, Air Force, Navy/Marine Corps)? [Note: Navy officers who work only with the Navy or Marine Corps, and Marine Corps officers who work only with the Marine Corp or Navy should answer NO to this question.]	
	○ Yes ○ No	If yes, are you assigned to a billet in that other Military Department? ○ Yes ○ No

◂Previous Next▸

For troubleshooting tips, click -->here
To email technical support, click here: jomhelp@lmi.org

Q-28	Do you typically serve full time with the armed forces of another nation or with an international military or treaty organization (e.g., a US officer assigned to a assignment in the headquarters of NATO; a liaison officer at the headquarters of a foreign military service; an officer assigned full time to an element of the United Nations, etc.)?
	○ Yes If yes, are you formally assigned to a billet in that organization? ○ Yes ○ No ○ No
Q-29	Are you assigned to both your own service and a joint, combined, or international organization? (Example, an officer assigned to a billet in the G3, Eighth US Army while simultaneously assigned to positions in the J3, US Forces Korea, and the C3, Combined Forces Command (ROK/US)). Such assignments are referred to as "Dual Hat Positions."
*Select **one** answer*	
	○ Yes ○ No ○ Unsure
Q-30	Do you serve in Joint Task Force Headquarters Staff?
*Select **one** answer*	
	○ Yes ○ No ○ Unsure
Q-31	Do you serve in a Joint Task Force Subordinate Organization?
*Select **one** answer*	
	○ Yes ○ No ○ Unsure

Q-32	Do you serve in a Joint Task Force Service Component?
	○ Yes If yes, are you permanently assigned to it? ○ Yes ○ No ○ No ○ Unsure

◂Previous Next▸

For troubleshooting tips, click -->here
To email technical support, click here: jomhelp@lmi.org

Q-33 Do you work in a Joint Program Office?

*Select **one** answer*

- ○ Yes
- ○ No
- ○ Unsure

Q-34 Where is the billet located?

*Select **one** answer*

- ○ United States (including Alaska and Hawaii)
- ○ Iraq
- ○ Other Middle East
- ○ South Asia (e.g., Pakistan, Afghanistan)
- ○ Korea
- ○ Cuba
- ○ Europe
- ○ Other nation outside of the U.S.
- ○ Afloat at sea

Q-35 Are you currently serving at your home base?

*Select **one** answer*

- ○ Yes
- ○ No
- ○ Unsure

Q-36 Are you currently receiving Family Separation Allowance (FSA), or would you collect FSA in this position if you had dependents?

*Select **one** answer*

- ○ Yes
- ○ No
- ○ Unsure

Q-37 Are you currently receiving special pay for duty subject to Hostile Fire or Imminent Danger?

*Select **one** answer*

- ○ Yes
- ○ No
- ○ Unsure

Q-38 Are you currently receiving special pay for Hardship Duty?

*Select **one** answer*

- ○ Yes
- ○ No
- ○ Unsure

Q-39 Is your pay subject to Combat Zone Tax Exclusion?

*Select **one** answer*

- ○ Yes
- ○ No
- ○ Unsure

Q-40 Please indicate the approximate percent of your work time you spend reviewing or deciding matters yourself, as opposed to preparing others to review or decide matters.

Reviewing / deciding matters myself	0 %
Preparing others to review/decide matters	0 %

Q-41 Indicate which one of the following statements best describes the primary focus of your efforts.

*Select **one** answer*

- ○ The primary focus of my efforts is on operational or supportability matters pertaining to a Combatant Commander's Area of Responsibility (AOR) or several AORs.
- ○ The primary focus of my efforts is on defense-wide issues or matters that affect one or more Combatant Commander's, Military Departments or Defense Agencies.
- ○ None of the above

Q-42 If you could choose only one of the following, which best summarizes the level of your job:

Please click on the hyperlinked choices if you are unsure

- ○ Strategic
- ○ Operational
- ○ Tactical

Q-43 On average, how many **HOURS** per week do you work?

*Please enter **whole** numbers only*

[ ]

Q-44 Select the tasks you typically perform.

*Please select **ALL** that apply.*

- ☐ Provide strategic direction and integration
- ☐ Legal affairs
- ☐ Inspector General activities
- ☐ Conduct mobilization
- ☐ Provide administrative or technical support
- ☐ Develop, conduct or provide intelligence, surveillance, and reconnaissance
- ☐ Provide or exercise command and control
- ☐ Employ forces
- ☐ Employ firepower or other assets
- ☐ Deploy and maneuver forces
- ☐ Provide or coordinate protection of the force, or protect the force
- ☐ Special operations
- ☐ Conduct deployment, redeployment, movement, or maneuver of forces
- ☐ Counter or manage deterrence of CBRNE weapons, or operate in a CBRNE environment
- ☐ Mapping, charting and geodesy
- ☐ Provide sustainment
- ☐ Provide logistics or combat service support
- ☐ Combat engineering
- ☐ Maintenance
- ☐ Industrial management
- ☐ Engineering
- ☐ Civil affairs and psychological operations
- ☐ Coordinate counter-proliferation in theater
- ☐ Foster multinational, interagency, alliance or regional relations
- ☐ Host nation security
- ☐ Targeting of enemy information systems
- ☐ Sustain theater forces communications, and computers (C4)
- ☐ Develop / assess joint doctrine
- ☐ Develop / assess joint policies
- ☐ Establish theater force requirements and readiness
- ☐ Resource / financial management
- ☐ Medical / health services
- ☐ Research, development, testing, evaluation & simulations
- ☐ Conduct force development
- ☐ Operations other than war
- ☐ Law enforcement
- ☐ Safety
- ☐ Acquisition/or Joint Program Management

◂Previous Next▸

For troubleshooting tips, click -->here

To email technical support, click here: jomhelp@lmi.org

Microsoft Internet Explorer

You are approximately half-way complete, by time required, for this census.

OK

Q-45	The tasks you chose in the previous question will appear below. Enter the number of **HOURS** per week you perform each task.

The tasks you chose in the previous question will appear below. Enter the number of HOURS per week you perform each task. (Total work hours answered in question 43 is 50. You may use the 'previous' button to edit the total hours or tasks)

Tasks	Hours
Provide strategic direction and integration	0

Q-46	The tasks you chose in the previous question will appear below. Next to each task is a percentage reflecting the percentage of time during the week you spend performing the task. This is based on your answer to the previous question, as well as your answer to question 43 regarding the total number of hours per week you work. Do the percentages accurately reflect how you spend your time during the week?

Change the values in question 45 to update the percentages.

Tasks	Percentage
Provide strategic direction and integration	NaN %

Q-47	For each of your identified tasks, please select the relative level of importance to your job.

Tasks	Level of Importance
Provide strategic direction and integration	-- Please Select --

Q-48	Please select the level of responsibility you hold for each of the tasks you perform

*Select **one** answer*

Tasks	Level of Responsibility
Provide strategic direction and integration	-- Please Select --

Q-49	With what organizations do you interact? For each organization that applies, please quantify the frequency of your interaction. ***Only select a frequency for those that apply.***

Organization	Frequency of likely Interaction
DOD - Office of the Secretary of Defense	Please Select If Applicable
DOD - Joint Chiefs of Staff	Please Select If Applicable
DOD - US Army	Please Select If Applicable
DOD - US Army National Guard	Please Select If Applicable
DOD - US Army Reserve	Please Select If Applicable
DOD - US Navy	Please Select If Applicable
DOD - US Naval Reserve	Please Select If Applicable
DOD - US Air Force	Please Select If Applicable
DOD - US Air National Guard	Please Select If Applicable
DOD - US Air Force Reserve	Please Select If Applicable
DOD - US Marine Corps	Please Select If Applicable
DOD - US Marine Corps Reserve	Please Select If Applicable
DOD - CENTCOM	Please Select If Applicable
DOD - EUCOM	Please Select If Applicable
DOD - JFCOM	Please Select If Applicable
DOD - NORTHCOM	Please Select If Applicable
DOD - PACOM	Please Select If Applicable
DOD - SOCOM	Please Select If Applicable
DOD - SOUTHCOM	Please Select If Applicable

DOD - Defense Finance and Accounting Service	Please Select If Applicable
DOD - Defense Legal Services Agency	Please Select If Applicable
DOD - Defense Information Systems Agency	Please Select If Applicable
DOD - DOD Computer Emergency Response Team	Please Select If Applicable
DHS – Bureau of Customs and Border Protection	Please Select If Applicable
DHS - Bureau of Immigration and Customs Enforcement	Please Select If Applicable
DHS - Emergency Preparedness & Response Directorate	Please Select If Applicable
DHS - Federal Emergency Management Agency	Please Select If Applicable
DHS - Federal Law Enforcement Training Center	Please Select If Applicable
DHS - Transportation Security Administration	Please Select If Applicable
DHS - US Coast Guard	Please Select If Applicable
DHS - US Secret Service	Please Select If Applicable
DHS - Other	Please Select If Applicable
Central Intelligence Agency	Please Select If Applicable
Other independent agency or government corporation	Please Select If Applicable
Executive Branch	Please Select If Applicable
Legislative Branch	Please Select If Applicable
Judicial Branch	Please Select If Applicable
Department of Agriculture	Please Select If Applicable
Department of Commerce	Please Select If Applicable
Department of Interior	Please Select If Applicable

Department of Justice	Please Select If Applicable
Department of State	Please Select If Applicable
Department of Transportation	Please Select If Applicable
Department of the Treasury	Please Select If Applicable
Department of Energy	Please Select If Applicable
Department of Health and Human Services	Please Select If Applicable
The United Nations	Please Select If Applicable
Treaty organizations (such as NATO)	Please Select If Applicable
US non-governmental orgs (such as The Red Cross)	Please Select If Applicable
Foreign non-governmental orgs (such as The Red Crescent)	Please Select If Applicable
Non-US military	Please Select If Applicable

◂Previous Next▸

For troubleshooting tips, click -->here
To email technical support, click here: jomhelp@lmi.org

Q-50	We would like to know with whom you interact. Please indicate this information by indicating the frequency of interaction.	
		Only select a frequency for those that apply.
	US Army personnel (Officer, Enlisted, or Civilian; Active duty, Nat'l Guard or Reserve)	Please Select If Applicable
	US Navy personnel (Officer, Enlisted or Civilian; Active duty or Reserve)	Please Select If Applicable
	US Air Force personnel (Officer, Enlisted or Civilian; Active duty, Nat'l Guard or Reserve)	Please Select If Applicable
	US Marine Corps personnel (Officer, Enlisted or Civilian; Active duty or Reserve)	Please Select If Applicable
	US Coast Guard Personnel (Officer, Enlisted or Civilian; Active duty or Reserve)	Please Select If Applicable
	Other DOD Civilian	Please Select If Applicable
	Other US Civilian	Please Select If Applicable
	Non-US Civilian	Please Select If Applicable
	Non-US Military Officer	Please Select If Applicable

◂Previous Next▸

For troubleshooting tips, click -->here
To email technical support, click here: jomhelp@lmi.org

Microsoft Internet Explorer

You are approximately three-quarters complete, by time required, for this census.

OK

Q-51 In this section, we would like to know two things:

a. What knowledge you feel was required or helpful to a person in this position, and
b. In what knowledge a person will gain – or can expect to gain – either familiarity or proficiency while in this position.

There are two sets of columns by each type of knowledge listed below. The first set of columns permits you to identify the type of knowledge that is either "required" or "helpful" for someone serving in this position. The second set of columns to identify those areas of expertise in which a person will gain either proficiency or familiarity while serving in this assignnment.

Please select buttons only from each subject row that applies to this position. It is likely that there will be lines or sections from which you do not select any buttons.

	For this position, I find this knowledge:		*My level of expertise at the end of this tour is expected to be:*		
National Military Capabilities, Organization and Command Structure	**Required**	**Helpful**	**Proficient**	**Familiar**	
Roles, relationships, and functions of the NCA, JCS, COCOMs, NSC, JFC, CJCS or the interagency process.	○	○	○	○	clear
Force structure requirements and resultant capabilities and limitations of US military forces	○	○	○	○	clear
How the US military plans, executes, and trains for joint, interagency, and multinational ops.	○	○	○	○	clear
Service-unique capability, limitation, doctrine, and command structure integration	○	○	○	○	clear

National Military Strategy	**Required**	**Helpful**	**Proficient**	**Familiar**	
Art and science of developing, deploying, employing, and sustaining the military resources	○	○	○	○	clear
Capabilities and limitations of the US force structure and their effect on joint military strategy	○	○	○	○	clear
Concepts of the strategic decision-making and defense planning processes	○	○	○	○	clear...
Resource needs, both national and international, for national defense	○	○	○	○	clear...
Key considerations that shape the development of national military strategy	○	○	○	○	clear...
Current National Military Strategy and other examples of US and foreign military strategies	○	○	○	○	clear...
DoD long term and immediate process for strategic planning and assessment	○	○	○	○	clear...
National Security Strategy	**Required**	**Helpful**	**Proficient**	**Familiar**	
National security policy process, to include the integration of the instruments of national power	○	○	○	○	clear...
Impact of defense acquisition and its implications for enhancing our joint military capabilities	○	○	○	○	clear...
Relationships between the military, Congress, NSC, DoD agencies, and the public	○	○	○	○	clear...
Developing, applying, and coordinating the instruments of national power	○	○	○	○	clear...
How national policy is turned into executable military strategies	○	○	○	○	clear...
Capabilities and vulnerabilities of US industry and infrastructure in a global market	○	○	○	○	clear...
National security technological environment for current and future competitive advantage	○	○	○	○	clear...

National Security Policy Process	Required	Helpful	Proficient	Familiar	
Origins, responsibilities, organization, and modus operandi of the NSC system.	○	○	○	○	clear...
How major governmental and NGO's influence and implement national security policies.	○	○	○	○	clear...
How the US government prioritizes among issues for developing national-level strategies	○	○	○	○	clear...
National Planning Systems and Processes	**Required**	**Helpful**	**Proficient**	**Familiar**	
National security decision-making system and the policy formulation process	○	○	○	○	clear...
Responsibilities and relationships of the interagency and the joint community, Responsibilities and relationships of the interagency and the joint community	○	○	○	○	clear...
DoD processes by which national ends, ways, and means are reconciled, integrated, and applied	○	○	○	○	clear...
How time, coordination, policy, politics, doctrine, and national power affect the planning process	○	○	○	○	clear...

Command, Control, Communications, Computers, INTELLIGENCE, SURVEILLANCE AND RECONNAISSANCE (C4ISR)	Required	Helpful	Proficient	Familiar	
How C4ISR systems apply at the tactical and operational levels of war	○	○	○	○	clear...
How IO is incorporated into both the deliberate and crisis action planning processes.	○	○	○	○	clear...
How opportunities and vulnerabilities are created by increased reliance on IT	○	○	○	○	clear...
Integrating IO and C4 to support the National Military and National Security Strategies	○	○	○	○	clear...
Integrating IO and C4 into the theater and strategic campaign development process	○	○	○	○	clear...
IO, IW, and C4I concepts in joint operations. , Role of the unified commander in developing theater plans, policies, and strategies	○	○	○	○	clear...
Theater Strategy and Campaigning	**Required**	**Helpful**	**Proficient**	**Familiar**	
Role of the unified commander in developing theater plans, policies, and strategies	○	○	○	○	clear...
Coordination of US military plans/actions with foreign forces, interagency and NGOs	○	○	○	○	clear...
How joint and multinational campaigns and operations support national objectives	○	○	○	○	clear...
Combatant Commander's perspective of the resources required to support campaign plans	○	○	○	○	clear...
Organization, responsibilities, and capabilities of military forces available to the JFCs, Organization, responsibilities, and capabilities of military forces available to the JFCs	○	○	○	○	clear...

Geo-Strategic Context	Required	Helpful	Proficient	Familiar	
Current social, cultural, political, economic, military, technological, and historical issues	○	○	○	○	clear...
Roles and influence of international organizations and other non-state actors	○	○	○	○	clear...
Key military, non-military, and transnational challenges to US national security	○	○	○	○	clear...
Instruments of National Power	**Required**	**Helpful**	**Proficient**	**Familiar**	
Fundamental characteristics, capabilities, and limitations instruments of national power	○	○	○	○	clear...
Employment of diplomatic, economic, military, and informational instruments of national power	○	○	○	○	clear...
Joint Operational Art	**Required**	**Helpful**	**Proficient**	**Familiar**	
Joint doctrine and the joint operational art	○	○	○	○	clear...
Integration of Service, joint, interagency, and multinational capabilities	○	○	○	○	clear...
Joint Warfare Fundamentals	**Required**	**Helpful**	**Proficient**	**Familiar**	
Each combatant command's mission, organization, and responsibilities	○	○	○	○	clear...
Joint aspects of military operations other than war (MOOTW)	○	○	○	○	clear...
Capabilities of other Services' weapon systems	○	○	○	○	clear...
Joint Campaigning	**Required**	**Helpful**	**Proficient**	**Familiar**	
JTF organization, including who can form a JTF and how and when a JTF is formed.	○	○	○	○	clear...
Characteristics of a joint campaign and the relationships of supporting capabilities	○	○	○	○	clear...

Joint Doctrine	Required	Helpful	Proficient	Familiar	
Current joint doctrine	○	○	○	○	clear...
Factors influencing joint doctrine	○	○	○	○	clear...
Relationship between Service doctrine and joint doctrine	○	○	○	○	clear...
Joint and Multinational Forces at the Operational Level of War	**Required**	**Helpful**	**Proficient**	**Familiar**	
Considerations for employing joint and multinational forces at the operational level of war	○	○	○	○	clear...
How theory and principles of war apply at the operational level of war	○	○	○	○	clear...
Relationships among national objectives, military objectives, and conflict termination	○	○	○	○	clear...
Relationships among the strategic, operational, and tactical levels of war	○	○	○	○	clear...
Joint Planning and Execution Processes	**Required**	**Helpful**	**Proficient**	**Familiar**	
Relationship between national objectives and means availability	○	○	○	○	clear...
Effect of time, coordination, policy changes, and political developments on the planning process	○	○	○	○	clear...
How national, joint, and Service intelligence organizations support JFCs	○	○	○	○	clear...
Integrating battle space support systems into campaign/theater planning and operations	○	○	○	○	clear...

Others	Required	Helpful	Proficient	Familiar	
Inspector General activities, legal/legislative, law enforcement, physical security or investigations	○	○	○	○	clear...
Special operations, operations other than war, tactical matters (i.e., training exercises, etc.)	○	○	○	○	clear...
Manpower / personnel, training, education, logistics, acquisition, or general administration	○	○	○	○	clear...
R&D, engineering, scientific matters (includes weather, environment, etc.), CBRNE matters	○	○	○	○	clear...
Medical or health services	○	○	○	○	clear...
Acquisition/Joint Program Management	○	○	○	○	clear...

◂Previous Next▸

For troubleshooting tips, click -->here
To email technical support, click here: jomhelp@lmi.org

Q-52 As a result of current events, my experience in this position was different from that of my predecessors.

*Select **one** answer*

- ○ Strongly Agree
- ○ Agree
- ○ Neither Agree nor Disagree
- ○ Disagree
- ○ Strongly Disagree
- ○ Not applicable

Q-53 My assessment of this position depends upon unusual current events, making it unlikely that future occupants will have the same experience.

*Select **one** answer*

- ○ Strongly Agree
- ○ Agree
- ○ Neither Agree nor Disagree
- ○ Disagree
- ○ Strongly Disagree
- ○ Not applicable

Q-54 This position gives me significant experience in multi-service matters.

*Select **one** answer*

- ○ Strongly Agree
- ○ Agree
- ○ Neither Agree nor Disagree
- ○ Disagree
- ○ Strongly Disagree

Q-55 This position gives me significant experience in multinational matters.

*Select **one** answer*

- ○ Strongly Agree
- ○ Agree
- ○ Neither Agree nor Disagree
- ○ Disagree
- ○ Strongly Disagree

Q-56 This position gives me significant experience in interagency matters.

*Select **one** answer*

- ○ Strongly Agree
- ○ Agree
- ○ Neither Agree nor Disagree
- ○ Disagree
- ○ Strongly Disagree

Q-57 In order to perform the duties of this position successfully, an individual would find JPME I.

*Select **one** answer*

- ○ Required
- ○ Desired
- ○ Not helpful
- ○ Not sure -- have no JPME experience

Q-58 In order to perform my duties successfully, I have found JPME II.

*Select **one** answer*

- ○ Required
- ○ Desired
- ○ Not helpful
- ○ Not sure -- have no JPME experience

Q-59 In order to perform my duties successfully, I have found joint training or education (other than JPME)

*Select **one** answer*

- ○ Required
- ○ Desired
- ○ Not helpful

Q-60 In order to perform my duties successfully, I have found prior experience in a joint environment

*Select **one** answer*

- ○ Required
- ○ Desired
- ○ Not helpful

Q-61 To what extent do you draw upon your primary specialty (i.e. AOC code, MOS, AFSC or Navy designator) to perform in this position?

*Select **one** answer*

- ○ Not at all
- ○ Some of the time
- ○ Half of the time
- ○ Most of the time
- ○ All of the time
- ○ Not sure

Q-62 To what extent do you draw upon knowledge of your Service's capabilities to perform in this position?

*Select **one** answer*

- ○ Not at all
- ○ Some of the time
- ○ Half of the time
- ○ Most of the time
- ○ All of the time
- ○ Not sure

◂Previous Next▸

For troubleshooting tips, click -->here
To email technical support, click here: jomhelp@lmi.org

Q-63 How many **MONTHS** did it take in this position to become comfortable in a joint environment?

*Select **one** answer*

- ○ [ ]
- ○ Not sure
- ○ Not in a joint environment
- ○ Not applicable for other reasons

Q-64 What is the planned length of your current assignment (in **MONTHS**)?

*Select **one** answer*

[ ]

Q-65 How many **MONTHS** do you think your assignment should last?

*Enter **whole** numbers only*

[ ]

Q-66 How many **MONTHS** do you think a typical permanent joint duty assignment should last?

*Select **one** answer*

[ ]

Q-67 How many **MONTHS** do you think a typical temporary (TAD/TDY) joint task force assignment should last? (Do not limit your answer by current tour length constraints, e.g.179 days)

*Select **one** answer*

[- Please select - ▾]

Q-68 Which of the following was most important to you in this assignment?

*Select **one** answer*

- ○ Service core competencies
- ○ Prior joint experience
- ○ Specialized training and orientation in joint matters
- ○ Functional expertise other than acquisition/joint program management (e.g. operations, intelligence, personnel, etc.)
- ○ Functional expertise in acquisition/joint program management
- ○ Other not listed here - please specify: [ ]

Q-69 In your opinion, what is the most important thing your successor should possess?

*Select **one** answer*

- ○ Service core competencies
- ○ Prior joint experience
- ○ Specialized training and orientation in joint matters
- ○ Functional expertise other than acquisition/joint program management (e.g. operations, intelligence, personnel, etc.)
- ○ Functional expertise in acquisition/joint program management
- ○ Other not listed here - please specify: [ ]

Q-70 A civilian could perform the duties and responsibilities of this position just as effectively.

*Select **one** answer*

- ○ Strongly Agree
- ○ Agree
- ○ Neither Agree nor Disagree
- ○ Disagree
- ○ Strongly Disagree

Q-71 My position requires unique knowledge of my own service and could not be performed by an officer of another service.

*Select **one** answer*

- ◯ Strongly Agree
- ◯ Agree
- ◯ Neither Agree nor Disagree
- ◯ Disagree
- ◯ Strongly Disagree
- ◯ Not applicable

Q-72 Morale problems will exist if joint duty credit is awarded for some positions in my immediate organization but not for others.

*Select **one** answer*

- ◯ Strongly Agree
- ◯ Agree
- ◯ Neither Agree nor Disagree
- ◯ Disagree
- ◯ Strongly Disagree
- ◯ Not applicable

Q-73 How much of your assessment of this billet is based on experience gained through TAD/TDY?

*Select **one** answer*

- ◯ Considerable amount
- ◯ Moderate amount
- ◯ Minimal amount
- ◯ Not at all
- ◯ Not applicable

Q-74 Have you served at least two months in a permanent joint duty billet AND, since September 2001, also served at least two months in a temporary joint task force (JTF) billet?

*Select **one** answer*

- ◯ Yes
- ◯ No

◂Previous Next▸

For troubleshooting tips, click -->here
To email technical support, click here: jomhelp@lmi.org

Q-75 In your experience, would you claim that, when comparing an equal period of time of time, you received an understanding of the joint environment more quickly while serving in the temporary joint task force (JTF) billet than while serving in the permanent joint duty assignment billet?

*Select **one** answer*

- ◯ YES, considerably more quickly
- ◯ YES, slightly more quickly
- ◯ About the same
- ◯ NO, more slowly
- ◯ NO, much more slowly

◂Previous Next▸

For troubleshooting tips, click -->here
To email technical support, click here: jomhelp@lmi.org

COMMENTS

Q-76 Is there anything else you would like to tell us?

Enter comments below.

◂Previous Submit

For troubleshooting tips, click -->here
To email technical support, click here: jomhelp@lmi.org

Microsoft Internet Explorer

Once you hit the 'OK' button, your answers will be recorded and the census will be complete.
If you have changes, hit 'Cancel' and use the 'previous' button to make any changes.

OK Cancel

Office of the Secretary of Defense

Thank you again for taking the time to participate in this critical endeavor. Again, please encourage others in your organization to take the time to provide their input, too. Your efforts will help us formulate appropriate legislative recommendations and make informed policy decisions to serve better the future joint officer and the Nation.

David S. C. Chu
Under Secretary of Defense
(Personnel and Readiness)

Census Complete. Thank You for Participating. You may now exit your browser.

APPENDIX B

Frequencies, Means, and Standard Deviations of Responses to the 2005 Joint Officer Management Census Survey Questions

Table B.1
Frequencies, Means, and Standard Deviations for the JOM Survey Questions[1, 2, 3]

Question Number	Survey Question	
		Percentage of Responses
Q-3	Are you the:	
	Select ***one*** *answer:*	
	Person occupying the billet?	80.8
	Supervisor of the billet?	6.1
	Another person designated to complete the survey?	13.1
	Number of missing responses	0
		Means (Standard Deviation)
Q-4	How many MONTHS have you been assigned to this billet?	
	Please ***enter*** *a* ***whole*** *number*	16.8 (13.4)
	Number of missing responses	43
		Percentage of Responses
Q-5	Is the billet you have been asked to evaluate a temporary JTF billet?	
	Select ***one*** *answer*	
	Yes	95.1
	No	4.9
	Number of missing responses	377
		Means (Standard Deviation)
Q-6	Please enter the total number of WEEKS you have spent on TDY/TAD (NOT including training and education during this assignment).	
	Please enter a ***whole*** *number*	9.5 (13.8)
	Number of missing responses	5,150

[1] For questions that asked respondents to select from among precoded answers, we present frequencies, i.e., percentages of officers who fall into particular categories. Where respondents were asked to fill in an answer (for example, the percentage of hours worked on a specific task), we present means and standard deviations for the respondents that answered that question.

[2] Questions Q-1 and Q-2 asked for identifying information to ensure that the respondent was eligible to take the survey.

[3] Sample size is 21,214.

Table B.1—Continued

Question Number	Survey Question	
Q-7	Please enter the total number of WEEKS you have spent on TDY/TAD due to training or education during this assignment.	
	*Please enter a **whole** number*	2.7 (5.5)
	Number of missing responses	5,158
Q-8	What is your pay grade?	
	Select one answer:	
	O-1	0.9
	O-2	3.3
	O-3	20.4
	O-4	28.0
	O-5	28.8
	O-6	12.0
	O-7	1.7
	O-8	1.1
	O-9	0.5
	O-10	0.2
	Civilian: GS-11	0.1
	Civilian: GS-12	0.5
	Civilian: GS-13	0.7
	Civilian: GS-14	0.8
	Civilian: GS-15	1.2
	Number of missing responses	705
	Questions 9 through 11 are optional. Answers will not be used for the JOM study.	
		Percentage of Responses
Q-9	What is your gender?	
	Select one answer:	
	Male	90.0
	Female	10.0
	Number of missing responses	1,712
Q-10	Are you Spanish/Hispanic/Latino?	
	Select one answer:	
	No, not Spanish/Hispanic/Latino	95.0
	Yes, Mexican, Mexican-American, Chicano, Puerto Rican, Cuban, or other Spanish/Hispanic/Latino	5.0
	Number of missing responses	1,931
Q-11	What is your race?	
	Mark one or more races to indicate what you consider yourself to be:	
	White	86.6
	Black or African-American	7.5
	Native-American or Alaska Native	1.2
	Asian (for example: Asian Indian, Chinese, Filipino, Japanese, Korean, or Vietnamese)	3.3
	Native Hawaiian or other Pacific Islander (for example: Samoan, Guamanian, Chamorro)	0.4
	Other	3.1
	Number of missing responses	1,986

Table B.1—Continued

Question Number	Survey Question	
Q-12	How many YEARS have you been a commissioned officer?	
	Please enter a ***whole*** *number*	15.8 (7.2)
	Number of missing responses	1,271
Q-13	Are you?	
	Select ***one*** *answer*	
	Active Duty List Officer	91.8
	Reserve Component Officer	4.7
	Retired/Separated (proxy cases only)	1.2
	Not applicable (proxy cases only)	2.2
	Number of missing responses	716
Q-14	Of the following list, what intermediate- or senior-level schools have you completed?	
	JOINT AND COMBINED SCHOOLS	
	National War College	8.1
	Industrial College of the Armed Forces	3.8
	School of Information Warfare and Strategy	0.5
	Joint Forces Staff College	18.5
	Joint and Combined Staff Officer School	3.3
	Joint and Combined Warfighting School	2.2
	U.S .ARMY SERVICE SCHOOLS	
	U.S. Army War College (resident)	4.8
	U.S. Army War College (nonresident)/Department of Distance Education	1.2
	Army Command and General Staff College (resident)	22.9
	Army Command and General Staff College (nonresident)	15.0
	U.S. NAVY SERVICE SCHOOLS	
	College of Naval Warfare	3.1
	College of Naval Command and Staff (resident)	5.0
	College of Distance Education (Navy intermediate-level college nonresident)	2.0
	U.S. AIR FORCE SERVICE SCHOOLS	
	Air War College (resident)	3.5
	Air War College (nonresident)	16.0
	Air Command and Staff College (resident)	11.8
	Air Command and Staff College (nonresident)	35.6
	U.S. MARINE CORPS SERVICE SCHOOLS	
	Marine Corps War College	0.3
	Marine Corps Command and Staff College	3.4
	Marine Corps College of Continuing Education (Marine Corps Command and Staff College nonresident)	3.3
	Number of missing responses	8,553

Table B.1—Continued

Question Number	Survey Question	
		Median Grade at Time of Completion
Q-14 (cont.)	Next to each school you have completed, note the paygrade at time of completion.	
	JOINT AND COMBINED SCHOOLS	
	National War College	0-4
	Industrial College of the Armed Forces	0-5
	School of Information Warfare and Strategy	0-5
	Joint Forces Staff College	0-4
	Joint and Combined Staff Officer School	0-4
	Joint and Combined Warfighting School	0-4
	U.S. ARMY SERVICE SCHOOLS	
	U.S. Army War College (resident)	0-5
	U.S. Army War College (nonresident)/Department of Distance Education	0-5
	Army Command and General Staff College (resident)	0-4
	Army Command and General Staff College (nonresident)	0-4
	U.S. NAVY SERVICE SCHOOLS	
	College of Naval Warfare	0-5
	College of Naval Command and Staff (resident)	0-4
	College of Distance Education (Navy intermediate-level college nonresident)	0-4
	U.S. AIR FORCE SERVICE SCHOOLS	
	Air War College (resident)	0-5
	Air War College (nonresident)	0-5
	Air Command and Staff College (resident)	0-4
	Air Command and Staff College (nonresident)	0-4
	U.S. MARINE CORPS SERVICE SCHOOLS	
	Marine Corps War College	0-5
	Marine Corps Command and Staff College	0-4
	Marine Corps College of Continuing Education (Marine Corps Command and Staff College nonresident)	0-4
Q-15	If you have received credit for post-graduate education at another civilian or military institution not listed above (to include international institutions), please enter the name of the institution and the paygrade you held when you graduated in the boxes below.	Not reported here
		Percentage of Responses
Q-16	Have you received credit for JPME I?	
	Select ***one*** *answer*	
	Yes	38.5
	No	49.3
	Not Sure	12.2
	Number of missing responses	734
Q-17	Have you received credit for JPME II?	
	Select ***one*** *answer*	
	Yes	16.6
	No	72.4
	Not Sure	11.0
	Number of missing responses	734

Table B.1—Continued

Question Number	Survey Question	
Q-18	What is your service and component?	
	*Select **one** answer*	
	U.S. Army	32.5
	U.S. Army National Guard	1.9
	U.S. Army Reserve	0.4
	U.S. Navy	19.6
	U.S. Naval Reserve	1.1
	U.S. Air Force	38.5
	U.S. Air National Guard	0.6
	U.S. Air Force Reserve	0.1
	U.S. Marine Corps	3.1
	U.S. Marine Corps Reserve	0.4
	Not applicable (proxy cases only)	1.9
	Number of missing responses	734
Q-19	Select your primary AFSC, e.g., 11MX (qualified Mobility Pilot) from the following list.	Not reported here
Q-20	Select your primary Area of Concentration (AOC) code, e.g., 11A (Infantry) from the following list.	Not reported here
Q-21	Select your 4-digit officer designated code, e.g., 111X (Surface Warfare Officer) from the following list.	Not reported here
Q-22	Select your 4-digit Military Occupational Specialty (MOS) code, e.g., 0802 (Field Artillery Officer).	Not reported here
Q-23	Are you a Joint Specialty Officer (JSO)?	
	Yes	10.2
	No	80.9
	Not Sure	9.0
	Number of missing responses	749
Q-24	Who is your 1st-level supervisor (e.g., Rater or Reporting Senior)?	
	U.S. Army	31.3
	U.S. Army National Guard	0.2
	U.S. Army Reserve	0.7
	U.S. Navy	16.2
	U.S. Naval Reserve	0.3
	U.S. Air Force	29.1
	U.S. Air National Guard	0.1
	U.S. Air Force Reserve	0.3
	U.S. Marine Corps	3.7
	U.S. Marine Corps Reserve	0.3
	DoD Civilian	15.7
	Other U.S. Civilian	0.8
	Non-U.S. Civilian	0.0
	Non-U.S. Military Officer	0.7
	Not applicable	0.6
	Number of missing responses	774

Table B.1—Continued

Question Number	Survey Question	
Q-25	Who is your 2nd-level supervisor (e.g., Intermediate Rater, Additional Rater, or Reviewing Senior)?	
	U.S. Army	22.5
	U.S. Army National Guard	0.2
	U.S. Army Reserve	0.3
	U.S. Navy	13.3
	U.S. Naval Reserve	0.1
	U.S. Air Force	26.3
	U.S. Air National Guard	0.1
	U.S. Air Force Reserve	0.2
	U.S. Marine Corps	3.1
	U.S. Marine Corps Reserve	0.2
	DoD Civilian	11.1
	Other U.S. Civilian	0.8
	Non-U.S. Civilian	0.1
	Non-U.S. Military Officer	0.5
	Not applicable	21.4
	Number of missing responses	774
Q-26	Who is your 3rd-level supervisor (e.g., Senior Rater or Reviewer)?	
	U.S. Army	29.3
	U.S. Army National Guard	0.2
	U.S. Army Reserve	0.4
	U.S. Navy	13.2
	U.S. Naval Reserve	0.1
	U.S. Air Force	27.9
	U.S. Air National Guard	0.2
	U.S. Air Force Reserve	0.2
	U.S. Marine Corps	4.2
	U.S. Marine Corps Reserve	0.1
	DoD Civilian	9.0
	Other U.S. Civilian	0.6
	Non-U.S. Civilian	0.0
	Non-U.S. Military Officer	0.2
	Not applicable	14.5
	Number of missing responses	774
Q-27	In your current position, do you serve full time with members from another Military Department (i.e., Army, Air Force, Navy/Marine Corps)? [Note: Navy officers who work only with the Navy or Marine Corps, and Marine Corps officers who work only with the Marine Corps or Navy, should answer NO to this question.]	
	Yes	61.2
	If yes, are you assigned to a billet in that other Military Department?	
	Yes	12.1
	No	87.9
	No	38.8
	Number of missing responses	774
Q-28	Do you typically serve full time with the armed forces of another nation or with an international military or treaty organization (e.g., a U.S. officer assigned to a assignment in the headquarters of NATO; a liaison officer at the headquarters of a foreign military service; an officer assigned full time to an element of the United Nations, etc.)?	
	Yes	11.2
	If yes, are you assigned to a billet in that other Military Department?	
	Yes	37.4
	No	62.6
	No	88.8
	Number of missing responses	787

Table B.1—Continued

Question Number	Survey Question	
Q-29	Are you assigned to both your own service and a joint, combined, or international organization? (Example, an officer assigned to a billet in the G3 (operations), Eighth U.S. Army while simultaneously assigned to positions in the J3 (operations), U.S. Forces Korea, and the C3 (operations), Combined Forces Command (ROK/US)). Such assignments are referred to as "Dual Hat Positions."	
	Yes	8.0
	No	88.4
	Unsure	3.6
	Number of missing responses	787
Q-30	Do you serve in Joint Task Force Headquarters Staff?	
	Yes	8.3
	No	89.0
	Unsure	2.7
	Number of missing responses	787
Q-31	Do you serve in a Joint Task Force Subordinate Organization?	
	Yes	6.1
	No	89.6
	Unsure	4.4
	Number of missing responses	787
Q-32	Do you serve in a Joint Task Force Service Component?	
	Yes	5.3
	If yes, are you permanently assigned to it?	
	Yes	49.4
	No	50.7
	No	91.1
	Unsure	3.6
	Number of missing responses	787
Q-33	Do you work in a Joint Program office?	
	Yes	5.2
	No	89.6
	Unsure	5.2
	Number of missing responses	808
Q-34	Where is the billet located?	
	United States (including Alaska and Hawaii)	79.9
	Iraq	2.4
	Other Middle East country	1.8
	South Asia (e.g., Pakistan, Afghanistan)	1.1
	Korea	3.4
	Cuba	0.0
	Europe	7.6
	Other nation outside of the United States	3.5
	Afloat at sea	0.3
	Number of missing responses	811
Q-35	Are you currently serving at your home base?	
	Yes	88.6
	No	10.1
	Unsure	1.3
	Number of missing responses	811

Table B.1—Continued

Question Number	Survey Question	
Q-36	Are you currently receiving Family Separation Allowance (FSA), or would you collect FSA in this position if you had dependents?	
	Yes	10.7
	No	87.8
	Unsure	1.5
	Number of missing responses	812
Q-37	Are you currently receiving special pay for duty subject to Hostile Fire or Imminent Danger?	
	Yes	10.5
	No	88.7
	Unsure	0.8
	Number of missing responses	812
Q-38	Are you currently receiving special pay for Hardship Duty?	
	Yes	8.0
	No	90.6
	Unsure	1.4
	Number of missing responses	812
Q-39	Is your pay subject to Combat Zone Tax Exclusion?	
	Yes	10.2
	No	88.8
	Unsure	1.0
	Number of missing responses	812
		Means (Standard Deviation)
Q-40	Please indicate the approximate percentage of your work time that you spend reviewing or deciding matters yourself, as opposed to preparing others to review or decide matters.	
	Reviewing/deciding matters myself	54.4 (26.8)
	Preparing others to review/decide matters	45.6 (26.8)
	Number of missing responses	810
		Percentage of Responses
Q-41	Indicate which one of the following statements best describes the primary focus of your efforts.	
	The primary focus of my efforts is on operational or supportability matters pertaining to a combatant commander's Area of Responsibility (AOR) or several AORs.	42.2
	The primary focus of my efforts is on defense-wide issues or matters that affect one or more combatant commander's, Military Departments, or Defense Agencies.	36.6
	None of the above	21.2
	Number of missing responses	818
Q-42	If you could choose only one of the following, which best summarizes the level of your job:	
	Strategic	39.1
	Operational	45.5
	Tactical	15.4
	Number of missing responses	812

Table B.1—Continued

Question Number	Survey Question	
		Means (Standard Deviation)
Q-43	On average, how many HOURS per week do you work? *Please enter **whole** numbers only*	52.6 (17.6)
	Number of missing responses	812
		Percentage of Responses
Q-44	Select the tasks you typically perform:	
	Provide strategic direction and integration	43.7
	Legal affairs	10.2
	Inspector General activities	4.7
	Conduct mobilization	8.0
	Provide administrative or technical support	51.1
	Develop, conduct, or provide intelligence, surveillance, and reconnaissance	19.7
	Provide or exercise command and control	28.5
	Employ forces	13.2
	Employ firepower or other assets	8.4
	Deploy and maneuver forces	11.0
	Provide or coordinate protection of the force, or protect the force	16.0
	Special operations	10.2
	Conduct deployment, redeployment, movement, or maneuver of forces	16.6
	Counter or manage deterrence of CBRNE weapons, or operate in a CBRNE environment	4.7
	Mapping, charting, and geodesy	4.3
	Provide sustainment	17.5
	Provide logistics or combat service support	19.3
	Combat engineering	1.4
	Maintenance	9.2
	Industrial management	3.5
	Engineering	9.0
	Civil affairs and psychological operations	5.0
	Coordinate counterproliferation in theater	2.5
	Foster multinational, interagency, alliance, or regional relations	21.2
	Host-nation security	5.6
	Targeting of enemy information systems	6.0
	Sustain theater forces communications and computers (C4)	6.8
	Develop assess joint doctrine	19.9
	Develop/assess joint policies	21.8
	Establish theater force requirements and readiness	9.9
	Resource/financial management	22.1
	Medical/health services	5.0
	Research, development, testing, evaluation, and simulations	17.0
	Conduct force development	9.0
	Operations other than war	15.4
	Law enforcement	3.0
	Safety	13.9
	Acquisition/or Joint Program Management	20.8
	Number missing all responses	812
Q-45	The tasks you chose in the previous question will appear below. Enter the number of HOURS per week you perform each task.	Not reported here

Table B.1—Continued

Question Number	Survey Question	
		Means (Standard Deviation) of Percentage of Time Spent by Those Who Perform a Given Task
Q-46	The tasks you chose in the previous question will appear below. Next to each task is a percentage reflecting the percentage of time during the week you spend performing the task. *This is based on your answer to the previous question, as well as your answer to question 43 regarding the total number of hours per week you work. Do the percentages accurately reflect how you spend your time during the week?*	
	Provide strategic direction and integration	27.1 (25.5)
	Legal affairs	14.3 (25.5)
	Inspector General activities	15.9 (28.2)
	Conduct mobilization	8.2 (11.3)
	Provide administrative or technical support	31.0 (28.1)
	Develop, conduct, or provide intelligence, surveillance, and reconnaissance	34.5 (33.4)
	Provide or exercise command and control	23.5 (24.1)
	Employ forces	11.7 (15.3)
	Employ firepower or other assets	13.5 (20.7)
	Deploy and maneuver forces	9.9 (12.6)
	Provide or coordinate protection of the force, or protect the force	9.9 (13.9)
	Special operations	23.2 (27.7)
	Conduct deployment, redeployment, movement, or maneuver of forces	12.3 (16.5)
	Counter or manage deterrence of CBRNE weapons, or operate in a CBRNE environment	13.1 (18.9)
	Mapping, charting, and geodesy	11.0 (16.9)
	Provide sustainment	13.4 (16.2)
	Provide logistics or combat service support	18.3 (21.9)
	Combat engineering	8.8 (15.0)
	Maintenance	10.1 (11.9)
	Industrial management	7.9 (10.6)
	Engineering	17.4 (20.9)
	Civil affairs and psychological operations	15.3 (23.2)
	Coordinate counterproliferation in theater	7.6 (9.8)
	Foster multinational, interagency, alliance, or regional relations	17.9 (19.7)
	Host-nation security	7.7 (10.2)
	Targeting of enemy information systems	12.2 (17.3)
	Sustain theater forces communications, and computers (C4)	22.4 (26.9)
	Develop/assess joint doctrine	12.2 (14.2)
	Develop/assess joint policies	13.0 (13.2)
	Establish theater force requirements and readiness	13.1 (16.2)
	Resource/financial management	15.2 (18.6)
	Medical/health services	20.0 (28.9)
	Research, development, testing, evaluation & simulations	20.9 (23.6)
	Conduct force development	16.3 (23.1)
	Operations other than war	15.8 (24.6)
	Law enforcement	7.7 (12.2)
	Safety	10.1 (18.2)
	Acquisition/or Joint Program Management	37.8 (30.7)

Table B.1—Continued

Question Number	Survey Question	Percentage Responding "of Primary or Vital Importance" (of those who perform a given task)
Q-47	For each of your identified tasks, please select the relative level of importance to your job:	
	Provide strategic direction and integration	74.8
	Legal affairs	28.6
	Inspector General activities	29.5
	Conduct mobilization	42.8
	Provide administrative or technical support	47.9
	Develop, conduct, or provide intelligence, surveillance, and reconnaissance	69.9
	Provide or exercise command and control	71.4
	Employ forces	67.9
	Employ firepower or other assets	73.5
	Deploy and maneuver forces	68.9
	Provide or coordinate protection of the force, or protect the force	54.1
	Special operations	65.4
	Conduct deployment, redeployment, movement, or maneuver of forces	58.4
	Counter or manage deterrence of CBRNE weapons, or operate in a CBRNE environment	48.7
	Mapping, charting, and geodesy	31.9
	Provide sustainment	54.9
	Provide logistics or combat service support	59.3
	Combat engineering	37.6
	Maintenance	48.4
	Industrial management	32.9
	Engineering	49.8
	Civil affairs and psychological operations	47.5
	Coordinate counterproliferation in theater	37.5
	Foster multinational, interagency, alliance, or regional relations	58.1
	Host-nation security	36.9
	Targeting of enemy information systems	51.4
	Sustain theater forces communications and computers (C4)	62.7
	Develop/assess joint doctrine	38.0
	Develop/assess joint policies	44.1
	Establish theater force requirements and readiness	53.2
	Resource financial management	50.6
	Medical/health services	49.3
	Research, development, testing, evaluation, and simulations	52.3
	Conduct force development	50.7
	Operations other than war	42.5
	Law enforcement	27.9
	Safety	53.6
	Acquisition/or Joint Program Management	71.4

Table B.1—Continued

Question Number	Survey Question	Percentage Responding "Mostly or Solely Responsible" (of those who perform a given task)
Q-48	Please select the level of responsibility you hold for each of the tasks you perform.	
	Provide strategic direction and integration	41.8
	Legal affairs	44.7
	Inspector General activities	39.3
	Conduct mobilization	32.4
	Provide administrative or technical support	46.1
	Develop, conduct, or provide intelligence, surveillance, and reconnaissance	40.6
	Provide or exercise command and control	53.0
	Employ forces	46.9
	Employ firepower or other assets	49.4
	Deploy and maneuver forces	46.9
	Provide or coordinate protection of the force, or protect the force	34.3
	Special operations	31.8
	Conduct deployment, redeployment, movement, or maneuver of forces	38.6
	Counter or manage deterrence of CBRNE weapons, or operate in a CBRNE environment	26.6
	Mapping, charting, and geodesy	25.8
	Provide sustainment	33.6
	Provide logistics or combat service support	37.9
	Combat engineering	34.7
	Maintenance	39.6
	Industrial management	24.6
	Engineering	31.2
	Civil affairs and psychological operations	33.2
	Coordinate counterproliferation in theater	15.7
	Foster multinational, interagency, alliance, or regional relations	30.0
	Host-nation security	16.0
	Targeting or enemy information systems	26.8
	Sustain theater forces communications and computers (C4)	32.4
	Develop/assess joint doctrine	19.2
	Develop/assess joint policies	21.5
	Establish theater force requirements and readiness	25.9
	Resource financial management	46.9
	Medical/health services	38.5
	Research, development, testing, evaluation, and simulations	35.4
	Conduct force development	31.9
	Operations other than war	26.4
	Law enforcement	26.6
	Safety	45.2
	Acquisition/or Joint Program Management	55.3

Table B.1—Continued

Question Number	Survey Question	Percentage of Responses
Q-49	With what organizations do you interact?	
	DOD—Office of the Secretary of Defense	51.1
	DOD—Joint Chiefs of Staff	48.0
	DOD—U.S. Army	61.4
	DOD—U.S. Army National Guard	32.1
	DOD—U.S. Army Reserve	31.3
	DOD—U..S Navy	56.0
	DOD—U.S. Naval Reserve	29.8
	DOD—U.S. Air Force	64.7
	DOD—U.S. Air National Guard	31.1
	DOD—U.S. Air Force Reserve	31.0
	DOD—U.S. Marine Corps	45.3
	DOD—U.S. Marine Corps Reserve	26.6
	DOD—CENTCOM	52.6
	DOD—EUCOM	44.4
	DOD—JFCOM	41.5
	DOD—NORTHCOM	38.9
	DOD—PACOM	46.5
	DOD—SOCOM	40.4
	DOD—SOUTHCOM	36.5
	DOD—SPACECOM	31.2
	DOD—STRATCOM	38.8
	DOD—TRANSCOM	35.8
	DOD—Industrial College of the Armed Forces	20.7
	DOD—Information Resource Management College	19.3
	DOD—Joint Forces Staff College	22.0
	DOD—Joint Military Intelligence College	19.9
	DOD—National Defense University	23.3
	DOD—Army Research Laboratory	20.8
	DOD—Defense Advanced Research Project Agency	23.3
	DOD—National Reconnaissance Office	27.8
	DOD—Defense Intelligence Agency	35.9
	DOD—Defense Logistics Agency	25.8
	DOD—Department of Defense Field Activities	20.2
	DOD—Defense Threat Reduction Agency	26.5
	DOD—Missile Defense Agency	25.0
	DOD—Defense Security Cooperation Agency	21.1
	DOD—National Geospatial-Intelligence Agency (formerly NIMA)	31.0
	DOD—National Security Agency—Central Security Service	30.2
	DOD—Defense Commissary Agency	18.9
	DOD—Defense Contract Audit Agency	19.8
	DOD—Defense Contract Management Agency	22.5
	DOD—Defense Finance and Accounting Service	23.9
	DOD—Defense Legal Services Agency	17.9
	DOD—Defense Information Systems Agency	25.4
	DOD—DOD Computer Emergency Response Team	18.7
	DHS—Bureau of Customs and Border Protection	19.7
	DHS—Bureau of Immigration and Customs Enforcement	19.2
	DHS—Emergency Preparedness and Response Directorate	18.0
	DHS—Federal Law Enforcement Training Center	20.3
	DHS—Transportation Security Administration	17.4
	DHS—U.S. Coast Guard	18.8
	DHS—U.S. Secret Service	26.4
	DHS—Other	19.5
	Central Intelligence Agency	20.0
	Other independent agency or government corporation	31.9
	Executive Branch	21.2
	Legislative Branch	20.4
	Judicial Branch	21.5
	Department of Agriculture	16.8
	Department of Commerce	16.9

Table B.1—Continued

Question Number	Survey Question	
Q-49 (cont.)	Department of Interior	17.4
	Department of Justice	16.8
	Department of State	19.5
	Department of Transportation	30.1
	Department of Treasury	18.3
	Department of Energy	17.4
	Department of Health and Human Services	19.3
	The United Nations	17.3
	Treaty organizations (such as NATO)	19.0
	U.S. nongovernmental organizations (such as the American Red Cross)	25.5
	Foreign non-governmental organizations (such as the Red Crescent)	21.5
	Non-U.S. military	18.3
	Number missing all responses	2,444
		Percentage Responding "Monthly," "Multiple Times Monthly," "Weekly," "Multiple Times Weekly," or "Daily"
Q-49 (cont.)	For each organization that applies, please quantify the frequency of your interaction. *Only select a frequency for those that apply.*	
	DOD—Office of the Secretary of Defense	27.7
	DOD—Joint Chiefs of Staff	27.1
	DOD—U.S. Army	42.3
	DOD—U.S. Army National Guard	9.8
	DOD—U.S. Army Reserve	9.5
	DOD—U.S. Navy	37.4
	DOD—U.S. Naval Reserve	9.0
	DOD—U.S. Air Force	49.0
	DOD—U.S. Air National Guard	9.6
	DOD—U.S. Air Force Reserve	9.7
	DOD—U.S. Marine Corps	26.0
	DOD—U.S. Marine Corps Reserve	5.3
	DOD—CENTCOM	30.4
	DOD—EUCOM	21.5
	DOD—JFCOM	18.7
	DOD—NORTHCOM	15.3
	DOD—PACOM	22.5
	DOD—SOCOM	17.7
	DOD—SOUTHCOM	12.8
	DOD—SPACECOM	7.9
	DOD—STRATCOM	16.0
	DOD—TRANSCOM	12.3
	DOD—Industrial College of the Armed Forces	0.9
	DOD—Information Resource Management College	0.6
	DOD—Joint Forces Staff College	1.2
	DOD—Joint Military Intelligence College	0.8
	DOD—National Defense University	2.3
	DOD—Army Research Laboratory	1.0
	DOD—Defense Advanced Research Project Agency	2.3
	DOD—National Reconnaissance Office	7.2
	DOD—Defense Intelligence Agency	16.4
	DOD—Defense Logistics Agency	5.9
	DOD—Department of Defense Field Activities	2.6
	DOD—Defense Threat Reduction Agency	6.0
	DOD—Missile Defense Agency	5.2
	DOD—Defense Security Cooperation Agency	3.3

Table B.1—Continued

Question Number	Survey Question	
Q-49 (cont.)	DOD—National Geospatial-Intelligence Agency (formerly NIMA)	11.0
	DOD—National Security Agency—Central Security Service	12.1
	DOD—Defense Commissary Agency	1.3
	DOD—Defense Contract Audit Agency	1.1
	DOD—Defense Contract Management Agency	3.7
	DOD—Defense Finance and Accounting Service	5.8
	DOD—Defense Legal Services Agency	0.8
	DOD—Defense Information Systems Agency	6.5
	DOD—DOD Computer Emergency Response Team	1.4
	DHS—Bureau of Customs and Border Protection	1.9
	DHS—Bureau of Immigration and Customs Enforcement	1.6
	DHS—Emergency Preparedness and Response Directorate	0.8
	DHS—FEMA	1.5
	DHS—Federal Law Enforcement Training Center	0.2
	DHS—Transportation Security Administration	1.5
	DHS—U.S. Coast Guard	6.5
	DHS—U.S. Secret Service	1.5
	DHS—Other	3.2
	Central Intelligence Agency	12.3
	Other independent agency or government corporation	4.9
	Executive Branch	3.5
	Legislative Branch	3.1
	Judicial Branch	0.3
	Department of Agriculture	0.4
	Department of Commerce	0.9
	Department of Interior	0.3
	Department of Justice	2.2
	Department of State	11.2
	Department of Transportation	1.3
	Department of Treasury	0.9
	Department of Energy	1.9
	Department of Health and Human Services	0.8
	The United Nations	1.7
	Treaty organizations (such as NATO)	6.9
	U.S. nongovernmental organizations (such as the American Red Cross)	2.9
	Foreign nongovernmental organizations (such as the Red Crescent)	1.2
	Non-U.S. military	22.6
	Number of missing responses	2,444
		Percentage of Responses
Q-50	We would like to know with whom you interact. Please indicate this information by indicating the frequency of interaction. *Only select a frequency for those that apply.*	
	U.S. Army personnel (Officer, Enlisted or Civilian, Active duty, National Guard, or Reserve)	86.0
	U.S. Navy personnel (Officer, Enlisted or Civilian, Active duty, or Reserve)	80.2
	U.S. Air Force personnel (Officer, Enlisted or Civilian, Active duty, National Guard, or Reserve)	87.9
	U.S. Marine Corps personnel (Officer, Enlisted or Civilian, Active duty, or Reserve)	73.2
	U.S. Coast Guard Personnel (Officer, Enlisted or Civilian, Active duty, or Reserve)	49.9

Table B.1—Continued

Question Number	Survey Question	
Q-50 (cont.)	Other DOD Civilian	87.0
	Other U.S. Civilian	71.2
	Non-U.S. Civilian	45.8
	Non-U.S. Military Officer	57.0
	Number of missing responses	1,036
		Percentage Responding "Monthly," "Multiple Times Monthly," "Weekly," "Multiple Times Weekly," or "Daily"
Q-50 (cont.)	We would like to know with whom you interact. Please indicate this information by indicating the frequency of interaction. *Only select a frequency for those that apply.*	
	U.S. Army personnel (Officer, Enlisted or Civilian, Active duty, Nat'l Guard, or Reserve)	75.7
	U.S. Navy personnel (Officer, Enlisted or Civilian, Active duty, or Reserve)	66.8
	U.S. Air Force personnel (Officer, Enlisted or Civilian, Active duty, National Guard, or Reserve)	79.8
	U.S. Marine Corps personnel (Officer, Enlisted or Civilian, Active duty, or Reserve)	55.7
	U.S. Coast Guard Personnel (Officer, Enlisted or Civilian, Active duty, or Reserve)	16.1
	Other DOD Civilian	81.0
	Other U.S. Civilian	62.6
	Non-U.S. Civilian	22.0
	Non-U.S. Military Officer	31.6
	Number of missing responses	1,036
		Percentage Responding "Required"/"Helpful"
Q-51	In this section, we would like to know two things: a. What knowledge do you feel was required or helpful to a person in this position, and b. What knowledge will a person gain or can expect to gain—either familiarity or proficiency—while in this position?	
	There are two sets of columns for each type of knowledge listed below. The first set of columns permits you to identify the type of knowledge that is either "required" or "helpful" for someone serving in this position. The second set of columns is to identify those areas of expertise in which a person will gain either proficiency or familiarity while serving in this assignment.	
	Please select buttons only from each subject row that applies to this position. It is likely that there will be lines or sections from which you do not select any buttons.	
	For this position, I find this knowledge required/helpful:	
	National Military Capabilities, Organization, and Command Structure	
	Roles, relationships, and functions of the NCA, JCS, COCOMs, NSC, JFC, CJCS or the *interagency process*	53.6/46.4
	Force structure requirements and resultant capabilities and limitations of U.S. military forces	46.2/53.8
	How the U.S. military plans, executes, and trains for joint, interagency, and multinational operations	50.2/49.8
	Service-unique capability, limitation, doctrine, and command structure integration	47.1/52.9

Table B.1—Continued

Question Number	Survey Question	
Q-51 (cont.)	**National Military Strategy**	
	Art and science of developing, deploying, employing, and sustaining the military resources	38.6/61.4
	Capabilities and limitations of the U.S. force structure and their effect on joint military strategy	38.1/61.9
	Concepts of the strategic decisionmaking and defense planning processes	39.5/60.5
	Resource needs, both national and international, for national defense	33.1/67.0
	Key considerations that shape the development of national military strategy	31.5/68.5
	Current National Military Strategy and other examples of U.S. and foreign military strategies	37.9/62.1
	DoD long-term and immediate process for strategic planning and assessment	32.9/67.1
	National Security Strategy	
	National security policy process, to include the integration of the instruments of national power	31.4/68.6
	Impact of defense acquisition and its implications for enhancing our joint military capabilities	32.8/67.2
	Relationships between the military, Congress, NSC, DoD agencies, and the public	39.7/60.3
	Developing, applying, and coordinating the instruments of national power	28.6/71.4
	How national policy is turned into executable military strategies	30.0/70.1
	Capabilities and vulnerabilities of U.S. industry and infrastructure in a global market	23.0/77.1
	National security technological environment for current and future competitive advantage	24.5/75.5
	National Security Policy Process	
	Origins, responsibilities, organization, and modus operandi of the NSC system	20.5/79.5
	How major government and NGOs influence and implement national security policies	20.7/79.3
	How the U.S. government prioritizes among issues for developing national-level strategies	20.9/79.1
	National Planning Systems and Processes	
	National security decisionmaking system and the policy foundation process	24.7/75.3
	Responsibilities and relationships of the interagency and the joint community	41.1/59.0
	DoD processes by which national ends, ways, and means are reconcile, integrated, and applied	25.8/74.2
	How time, coordination, policy, politics, doctrine, and national power affect the planning process	29.7/70.4
	Command, Control, Communications, Computers, Intelligence, Surveillance, and Reconnaissance (C4ISR)	
	How C4ISR systems apply at the tactical and operational levels of war	45.0/55.0
	How IO is incorporated into both the deliberate and crisis action planning processes	37.8/62.2
	How opportunities and vulnerabilities are created by increased reliance on IT	35.5/64.5
	Integrating IQ and C4 to support the National Military and National Security Strategies	32.4/67.6
	Integrating IO and C4 into the theater and strategic campaign development process	34.0/66.0
	IQ, IW, and C4I concepts in joint operations	35.3/64.7

Table B.1—Continued

Question Number	Survey Question	
Q-51 (cont.)	**Theater Strategy and Campaigning**	
	Role of the unified commander in developing theater plans, policies, and strategies	42.5/57.6
	Coordination of U..S. military plans/actions with foreign forces, interagency organizations and NGOs	39.5/60.5
	How joint and multinational campaigns and operations support national objectives	39.0/70.0
	Combatant commander's perspective of the resources required to support campaign plans	41.2/58.8
	Organization, responsibilities, and capabilities of military forces available to the JFCs	42.3/57.7
	Geo-Strategic Context	
	Current social, cultural, political, economic, military, technological, and historical issues	37.0/63.1
	Roles and influence of international organizations and other non-state actors	31.4/68.6
	Key military, non-military, and transnational challenges to U.S. national security	35.6/64.5
	Instruments of National Power	
	Fundamental characteristics, capabilities, and limitations of instruments of national power	31.3/68.7
	Employment of diplomatic, economic, military, and informational instruments and national power	32.1/67.9
	Joint Operational Art	
	Joint doctrine and the joint operational art	40.5/59.5
	Integration of Service, joint, interagency, and multinational capabilities	45.0/55.1
	Joint Warfare Fundamentals	
	Each combatant command's mission, organization, and responsibilities	41.2/58.8
	Joint aspects of military operations other than war (MOOTW)	35.5/64.6
	Capabilities of other Services' weapon systems	31.7/68.3
	Joint Campaigning	
	JTF organization, including who can form a JTF and how and when a JTF is formed	36.2/63.8
	Characteristics of a joint campaign and the relationships of supporting capabilities	36.3/63.7
	Joint Doctrine	
	Current joint doctrine	42.2/57.8
	Factors influencing joint doctrine	33.0/67.0
	Relationship between Service doctrine and joint doctrine	37.4/62.6
	Joint and Multinational Forces at the Operational Level of War	
	Considerations for employing joint and multinational forces at the operational level of war	33.3/66.7
	How theory and principles of war apply at the operational level of war	30.1/69.9
	Relationships among national objectives, military objectives, and conflict termination	32.4/67.6
	Relationships among the strategic, operational, and tactical levels of war	40.5/59.5

Table B.1—Continued

Question Number	Survey Question	
Q-51 (cont.)	**Joint Planning and Execution Processes**	
	Relationship between national objectives and means availability	31.7/68.3
	Effect of time, coordination, policy changes, and political developments on the planning process	32.3/67.7
	How national, joint, and Service intelligence organizations support JFCs	34.2/65.8
	Integrating battle space support systems into campaign/theater planning and operations	30.5/69.5
	Others	
	Inspector General activities, legal/legislative, law enforcement, physical security or investigations	20.5/79.5
	Special operations, operations other than war, tactical matters (i.e., training exercises, etc.)	33.9/66.1
	Manpower/personnel, training, education, logistics, acquisition, or general administration	40.0/60.0
	R&D, engineering, scientific matters (includes weather, environment, etc.), CBRNE matters	76.3/23.7
	Medical or health services	4.1/85.9
	Acquisition/Joint Program Management	31.3/68.7
	Number missing all responses	1,669
		Percentage Responding "Proficient"/"Familiar"
Q-51 (cont.)	In this section, we would like to know two things: a. What knowledge do you feel was required or helpful to a person in this position, and b. What knowledge will a person gain or can expect to gain—either familiarity or proficiency—while in this position? *There are two sets of columns for each type of knowledge listed below. The first set of columns permits you to identify the type of knowledge that is either "required" or "helpful" for someone serving in this position. The second set of columns is to identify those areas of expertise in which a person will gain either proficiency or familiarity while serving in this assignment.* *Please select buttons only from each subject row that applies to this position. It is likely that there will be lines or sections from which you do not select any buttons.* My level of expertise at the end of this tour is expected to be proficient/familiar with:	
	National Military Capabilities, Organization, and Command Structure	
	Roles, relationships, and functions of the NCA, JCS, COCOMs, NSC, JFC, CJCS ,or the interagency process	51.7/48.3
	Force structure requirements and resultant capabilities and limitations of U.S. military forces	45.8/54.2
	How the U.S. military plans, executes, and joint, interagency, and multinational operations	47.6/52.4
	Service-unique capability, limitation, doctrine, and command structure integration	45.0/55.0

Table B.1—Continued

Question Number	Survey Question	
Q-51 (cont.)	**National Military Strategy**	
	Art and science of developing, deploying, employing, and sustaining the military resources	38.5/61.5
	Capabilities and limitations of the U.S. force structure and their effect on joint military strategy	36.6/63.4
	Concepts of the strategic decisionmaking and defense planning processes	38.1/61.9
	Resource needs, both national and international, for national defense	31.6/68.4
	Key considerations that shape the development of national military strategy	30.4/69.6
	Current National Military Strategy and other examples of U.S. and foreign military strategies	35.4/64.6
	DoD long-term and immediate process for strategic planning and assessment	30.3/69.7
	National Security Strategy	
	National security policy process, to include the integration of the instruments of national power	31.2/68.8
	Impact of defense acquisition and its implications for enhancing our joint military capabilities	31.7/68.3
	Relationships between the military, Congress, NSC, DoD agencies, and the public	39.0/61.0
	Developing, applying, and coordinating the instruments of national power	28.9/71.1
	How national policy is turned into executable military strategies	30.3/69.7
	Capabilities and vulnerabilities of U.S. industry and infrastructure in a global market	22.4/77.6
	National security technological environment for current and future competitive advantage	23.3/76.7
	National Security Policy Process	
	Origins, responsibilities, organization, and modus operandi of the NSC system	19.8/80.2
	How major government and NGO's influence and implement national security policies	20.4/79.6
	How the U.S. government prioritizes among issues for developing national-level strategies	20.2/79.8
	National Planning Systems and Processes	
	National security decisionmaking system and the policy foundation process	23.8/76.2
	Responsibilities and relationships of the interagency and the joint community	37.6/62.5
	DoD processes by which national ends, ways, and means are reconcile, integrated, and applied	25.5/75.6
	How time, coordination, policy, politics, doctrine, and national power affect the planning process	29.3/70.8
	Command, Control, Communications, Computers, Intelligence, Surveillance, and Reconnaissance (C41SR)	
	How C41SR systems apply at the tactical and operational levels of war	42.1/57.9
	How IQ is incorporated into both the deliberate and crisis action planning processes	34.8/65.2
	How opportunities and vulnerabilities are created by increased reliance on IT	33.8/66.3
	Integrating IQ and C4 to support the National Military and National Security Strategies	30.0/70.0
	Integrating IA and C4 into the theater and strategic campaign development process	31.3/68.7
	IQ, IW, and C41 concepts in joint operations	32.3/67.8

Table B.1—Continued

Question Number	Survey Question	
Q-51 (cont.)	**Theater Strategy and Campaigning**	
	Role of the unified commander in developing theater plans, policies, and strategies	39.7/60.3
	Coordination of U.S. military plans/actions with foreign forces, interagency organizations and NGOs	37.2/62.9
	How joint and multinational campaigns and operations support national objectives	37.7/62.3
	Combatant commander's perspective of the resources required to support campaign plans	38.4/61.6
	Organization, responsibilities, and capabilities of military forces available to the JFCs	39.5/60.5
	Geo-Strategic Context	
	Current social, cultural, political, economic, military, technological, and historical issues	36.6/63.4
	Roles and influence of international organizations and other non-state actors	30.7/69.3
	Key military, non-military, and transnational challenges to U.S. national security	34.0/66.0
	Instruments of National Power	
	Fundamental characteristics, capabilities, and limitations on instruments of national power	31.2/68.8
	Employment of diplomatic, economic, military, and informational instruments of national power	32.0/68.0
	Joint Operational Art	
	Joint doctrine and the joint operational art	37.8/62.2
	Integration of Service, joint, interagency, and multinational capabilities	41.6/58.4
	Joint Warfare Fundamentals	
	Each combatant command's mission, organization, and responsibilities	39.7/60.3
	Joint aspects of military operations other than war (MOOTW)	33.8/66.2
	Capabilities of other Services' weapon systems	30.3/69.7
	Joint Campaigning	
	JTF organization, including who can form a JTF and how and when a JTF is formed	35.2/64.9
	Characteristics of a joint campaign and the relationships of supporting capabilities	34.9/65.1
	Joint Doctrine	
	Current joint doctrine	37.0/63.0
	Factors influencing joint doctrine	31.3/68.7
	Relationship between Service doctrine and joint doctrine	33.5/66.5
	Joint and Multinational Forces at the Operational Level of War	
	Considerations for employing joint and multinational forces at the operational level of war	31.7/68.3
	How theory and principles of war apply at the operational level of war	30.8/69.2
	Relationships among national objectives, military objectives, and conflict termination	32.0/68.0
	Relationships among the strategic, operational, and tactical levels of war	40.4/59.6

Table B.1—Continued

Question Number	Survey Question	
Q-51	**Joint Planning and Execution Processes**	
	Relationship between national objectives and means availability	30.0/70.0
	Effect of time, coordination, policy changes, and political developments on the planning process	31.0/69.1
	How national, joint, and Service intelligence organizations support JFCs	32.0/68.0
	Integrating battle space support systems into campaign/theater planning and operations	28.2/71.8
	Others	
	Inspector General activities, legal/legislative, law enforcement, physical security or investigations	20.5/79.5
	Special operations, operations other than war, tactical matters (i.e., training exercises, etc.)	32.5/67.6
	Manpower/personnel, training, education, logistics, acquisition, or general administration	39.4/60.6
	R&D, engineering, scientific matters (includes weather, environment, etc.), CBRNE matters	23.3/76.8
	Medical or health services	14.0/86.0
	Acquisition/Joint Program Management	31.1/68.9
	Number missing all responses	1,669
		Percentage of Responses
Q-52	As a result of current events, my experience in this position was different from that of my predecessors.	
	*Select **one** answer*	
	Strongly Agree	20.5
	Agree	29.0
	Neither Agree nor Disagree	24.5
	Disagree	16.1
	Strongly Disagree	3.8
	Not applicable	6.1
	Number of missing responses	958
Q-53	My assessment of this position depends upon unusual current events, making it unlikely that future occupants will have the same experience.	
	Select ***one*** answer	
	Strongly Agree	5.7
	Agree	17.6
	Neither Agree nor Disagree	25.8
	Disagree	36.5
	Strongly Disagree	11.4
	Not applicable	3.0
	Number of missing responses	958
Q-54	This position gives me significant experience in multiservice matters.	
	Select ***one*** answer	
	Strongly Agree	29.5
	Agree	36.9
	Neither Agree nor Disagree	16.7
	Disagree	13.3
	Strongly Disagree	3.6
	Number of missing responses	958

Table B.1—Continued

Question Number	Survey Question	Percentage Responding "Required"/"Helpful"[a]
Q-55	This position gives me significant experience in multinational matters.	
	*Select **one** answer*	
	Strongly Agree	21.1
	Agree	29.6
	Neither Agree nor Disagree	21.1
	Disagree	21.4
	Strongly Disagree	6.9
	Number of missing responses	958
Q-56	This position gives me significant experience in interagency matters.	
	*Select **one** answer*	
	Strongly Agree	23.5
	Agree	36.3
	Neither Agree nor Disagree	20.9
	Disagree	15.1
	Strongly Disagree	4.2
	Number of missing responses	958
Q-57	In order to perform the duties of this position successfully, an individual would find JPME I	
	*Select **one** answer*	
	Required	24.6
	Desired	31.0
	Not helpful	7.8
	Not sure—have no JPME experience	36.6
	Number of missing responses	958
Q-58	In order to perform my duties successfully, I have found JPME II	
	*Select **one** answer*	
	Required	11.6
	Desired	29.9
	Not helpful	8.9
	Not sure—have no JPME experience	49.6
	Number of missing responses	958
Q-59	In order to perform my duties successfully, I have found joint training or education (other than JPME)	
	*Select **one** answer*	
	Required	18.2
	Desired	60.8
	Not helpful	21.0
	Number of missing responses	958
Q-60	In order to perform my duties successfully, I have found prior experience in a joint environment	
	*Select **one** answer*	
	Required	13.8
	Desired	64.8
	Not helpful	21.4
	Number of missing responses	959

Table B.1—Continued

Question Number	Survey Question	
Q-61	To what extent do you draw upon your primary specialty (i.e., AOC code, MOS, AFSC, or Navy designator) to perform in this position?	
	*Select **one** answer*	
	Not at all	6.2
	Some of the time	23.7
	Half of the time	10.1
	Most of the time	29.1
	All of the time	28.6
	Not sure	2.4
	Number of missing responses	960
Q-62	To what extent do you draw upon knowledge of your service's capabilities to perform in this position?	
	*Select **one** answer*	
	Not at all	4.6
	Some of the time	33.3
	Half of the time	12.2
	Most of the time	29.1
	All of the time	18.0
	Not sure	2.8
	Number of missing responses	960
Q-63	How many MONTHS did it take in this position to become comfortable in a joint environment?	
	*Select **one** answer*	
	Answered	60.2
	Not sure	12.0
	Not in a joint environment	21.2
	Not applicable for other reasons	6.6
	Number of missing responses	979
		Means (Standard Deviation)
Q-63	How many MONTHS did it take in this position to become comfortable in a joint environment?	5.1 (4.4)
Q-64	What is the planned length of your current assignment (in MONTHS)? *Enter **whole** numbers only*	29.5 (11.4)
Q-65	How many MONTHS do you think your assignment should last? *Enter **whole** numbers only*	29.9 (12.0)
Q-66	How many MONTHS do you think a typical permanent joint duty assignment should last? *Enter **whole** numbers only*	30.2 (9.7)
Q-67	How many MONTHS do you think a typical temporary (TAD/TDY) joint task force assignment should last? (Do not limit your answer by current tour length constraints, e.g., 179 days.)	7.0 (3.4)

Table B.1—Continued

Question Number	Survey Question	Percentage of Responses
Q-68	Which of the following was not important to you in this assignment?	
	Select ***one*** *answer*	
	Service core competencies	29.1
	Prior joint experience	11.7
	Specialized training and orientation in joint matters	9.7
	Functional expertise other than acquisition/joint program management (e.g., operations, intelligence, personnel, etc.)	25.9
	Functional expertise in acquisition/joint program management	12.2
	Other not listed here—please specify:	11.4
	Number of missing responses	979
Q-69	In your opinion, what is the most important thing your successor should possess?	
	Select ***one*** *answer*	
	Service core competencies	27.4
	Prior joint experience	9.9
	Specialized training and orientation in joint matters	11.3
	Functional expertise other than acquisition/joint program management (e.g., operations, intelligence, personnel, etc.)	25.6
	Functional expertise in acquisition/joint program management	13.8
	Other not listed here—please specify:	12.1
	Number of missing responses	4,889
Q-70	A civilian could perform the duties and responsibilities of this position just as effectively.	
	Select ***one*** *answer*	
	Strongly Agree	10.3
	Agree	19.6
	Neither Agree nor Disagree	13.7
	Disagree	26.2
	Strongly Disagree	30.2
	Number of missing responses	979
Q-71	My position is requires unique knowledge of my own service and could not be performed by an officer of another service.	
	Select ***one*** *answer*	
	Strongly Agree	19.5
	Agree	17.8
	Neither Agree nor Disagree	13.4
	Disagree	30.6
	Strongly Disagree	18.1
	Not applicable	0.7
	Number of missing responses	4,887
Q-72	Morale problems will exist if joint duty credit is awarded for some positions in my immediate organization but not for others.	
	Select ***one*** *answer*	
	Strongly Agree	21.4
	Agree	28.3
	Neither Agree nor Disagree	23.6
	Disagree	13.9
	Strongly Disagree	4.9
	Not applicable	7.9
	Number of missing responses	4,887

Table B.1—Continued

Question Number	Survey Question	
Q-73	How much of your assessment of this billet is based on experience gained through TAD/TDY?	
	*Select **one** answer*	
	Considerable amount	11.9
	Moderate amount	18.9
	Minimal amount	29.9
	Not at all	29.0
	Not applicable	10.3
	Number of missing responses	4,887
Q-74	Have you served at least two months in a permanent joint duty billet AND, since September 2001, also served at least two months in a temporary joint task force (JTF) billet?	
	*Select **one** answer*	
	Yes	17.4
	No	82.6
	Number of missing responses	4,887
Q-75	In your experience, would you claim that, when comparing an equal period of time, you received an understanding of the joint environment more quickly while serving in the temporary joint task force (JTF) billet than while serving in the permanent joint duty assignment billet?	
	Select **one** *answer*	
	YES, considerably more quickly	22.8
	YES, slightly more quickly	23.9
	About the same	38.2
	NO, more slowly	12.1
	NO, much more slowly	3.0
	Number of missing responses	18,379

APPENDIX C

Supporting Tables for Chapter Six on Required or Helpful Types of Knowledge and Experience

Table C.1
General Knowledge Categories and Specific Types of Knowledge Within the Categories[1]

National Military Capabilities, Organization, and Command Structure
Roles, relationships, and functions of the NCA, JCS, COCOMs, NSC, JFC, CJCS, or the interagency process
Force structure requirements and resultant capabilities and limitations of U.S. military forces
How the U.S. military plans, executes, and trains for joint, interagency, multinational operations
Service-unique capability, limitation, doctrine, and command structure integration

National Military Strategy
Art and science of developing, deploying, employing, and sustaining military resources
Capabilities and limitations of the U.S. force structure and their effect on joint military strategy
Concepts of the strategic decisionmaking and defense planning processes
Resource needs, both national and international, for national defense
Key considerations that shape the development of national military strategy
Current National Military Strategy and other examples of U.S. and foreign military strategies
DoD long-term and immediate processes for strategic planning and assessment

National Security Strategy
The national security policy process, to include the integration of national instruments of power
The impact of defense acquisition and its implications for enhancing joint military capabilities
Relationships between the military, Congress, NSC, DoD agencies, and the public
Developing, applying, and coordinating the instruments of national power
How national policy is turned into executable military strategies
Capabilities and vulnerabilities of U.S. industry and infrastructure in a global market
The national security technological environment for current and future competitive advantage

National Security Policy Process
Origins, responsibilities, organization, and modus operandi of the NSC system
How major governmental and NGOs influence and implement national security policies
How the U.S. government prioritizes among issues for developing national-level strategies

National Planning Systems and Processes
National security decisionmaking and the policy formulation process
Responsibilities and relationships of the interagency and joint community
DoD processes by which national ends, ways, and means are reconciled, integrated, and applied
How time, coordination, policy, politics, doctrine, and national power affect the planning process

Command, Control, Communications, Computers, Intelligence, Surveillance, and Reconnaissance (C4ISR)
How C4ISR systems apply at the tactical and operational levels of war
How IO is incorporated into both deliberate and crisis action-planning processes

[1] The descriptions of the general and specific knowledge categories listed in this appendix are verbatim from the JOM census survey.

Table C.1—Continued

Command, Control, Communications, Computers, Intelligence, Surveillance, and Reconnaissance (C4ISR) (cont.)
How opportunities and vulnerabilities are created by increased reliance on IT
Integrating IO and C4 to support National Military and National Security Strategies
Integrating IO and C4 into the theater and strategic campaign development process
IO, IW, C4I concepts in joint operations
Theater Strategy and Campaigning
Role of the unified commander in developing theater plans, policies, and strategies
Coordination of U.S. military plans/actions with foreign forces, interagency, and NGOs
How joint and multinational campaigns and operations support national objectives
Combatant commander's perspective of the resources required to support campaign plans
Organization, responsibilities, and capabilities of military forces available to the JFCs
Geo-Strategic Context
Current social, cultural, political, economic, military, technological, and historical issues
Roles and influence of international organizations and other nonstate actors
Key military, nonmilitary, and transnational challenges to U.S. national security
Instruments of National Power
Fundamental characteristics, capabilities, and limitations of instruments of national power
Employing diplomatic, economic, military, and informational instruments of national power
Joint Operation Art
Joint doctrine and the joint operational art
Integration of service, joint, interagency, and multinational capabilities
Joint Warfare Fundamentals
Each combatant command's mission, organization, and responsibilities
Joint aspects of military operations other than war
Capabilities of other services' weapons systems
Joint Campaigning
JTF organization, including who can form a JTF and how and when a JTF is formed
Characteristics of a joint campaign and the relationships of supporting capabilities
Joint Doctrine
Current joint doctrine
Factors influencing joint doctrine
Relationship between service doctrine and joint doctrine
Joint and Multinational Forces at the Operational Level of War
Considerations for employing joint and multinational forces at the operational level of war
How theory and principles of war apply at the operational level of war
Relationships among national objectives, military objectives, and conflict termination
Relationships among the strategic, operational, and tactical levels of war
Joint Planning and Execution Processes
Relationships between national objectives and means availability
Effect of time, coordination, policy changes, and political developments on the planning process
How national, joint, and service intelligence organizations support JFCs
Integrating battlespace support systems into campaign/theater planning and operations
Other Knowledge Types
Inspector General activities, legal/legislative, law enforcement, physical security, or investigations
Special operations, operations other than war, or tactical matters (i.e., training exercises, etc.)
Manpower/personnel, training, education, logistics, acquisition, or general administration
Research and development, engineering, scientific matters (includes weather, environment, etc.), CBRNE matters
Medical or health services
Acquisition/Joint Program Management

Table C.2
Percentage of Officers Reporting that Specific Types of Knowledge Are Required or Helpful

Broad Knowledge Categories	Specific Knowledge Categories	Required	Required or Helpful
National Military Capabilities, Organization, and Command Structure	Roles, relationships, and functions of the NCA, JCS, COCOMs, NSC, JFC, CJCS, or the interagency process	45.0	84.0
	Force structure requirements and resultant capabilities and limitations of U.S. military forces	37.4	81.0
	How the U.S. military plans, executes, and trains for joint, interagency, multinational operations	41.4	82.6
	Service-unique capability, limitation, doctrine, and command structure integration	39.1	83.0
National Military Strategy	Art and science of developing, deploying, employing, and sustaining military resources	29.3	76.0
	Capabilities and limitations of the U.S. force structure and their effect on joint military strategy	28.0	73.5
	Concepts of the strategic decisionmaking and defense planning processes	28.5	72.3
	Resource needs, both national and international, for national defense	23.7	71.7
	Key considerations that shape the development of national military strategy	21.5	68.3
	Current National Military Strategy and other examples of U.S and foreign military strategies	26.7	70.5
	DoD long-term and immediate process for strategic planning and assessment	22.6	68.8
	The national security policy process, to include the integration of national instruments of power	21.1	67.3
	The impact of defense acquisition and its implications for enhancing joint military capabilities	22.7	69.1
	Relationships between the military, Congress, NSC, DoD agencies, and the public	29.0	72.9
	Developing, applying, and coordinating the instruments of national power	18.4	64.3
	How national policy is turned into executable military strategies	19.4	64.8
	Capabilities and vulnerabilities of U.S. industry and infrastructure in a global market	14.4	62.9
	The national security technological environment for current and future competitive advantage	15.2	62.2
National Security Policy Process	Origins, responsibilities, organization, and modus operandi of the NSC system	12.4	60.6
	How major governmental and NGOs influence and implement national security policies	12.7	61.3
	How the U.S. government prioritizes among issues for developing national-level strategies	12.6	60.3

Table C.2—Continued

Broad Knowledge Categories	Specific Knowledge Categories	Required	Required or Helpful
National Planning Systems and Processes	National security decisionmaking and the policy formulation process	15.3	62.1
	Responsibilities and relationships of the interagency and joint community	28.0	68.2
	DoD processes by which national ends, ways, and means are reconciled, integrated, and applied	15.8	61.3
	How time, coordination, policy, politics, doctrine, and national power affect the planning process	18.9	63.6
Command, Control, Communications, Computers, Intelligence, Surveillance, and Reconnaissance (C4ISR)	How C4ISR systems apply at the tactical and operational levels of war	33.3	73.9
	How IO is incorporated into both deliberate and crisis action-planning processes	26.1	69.0
	How opportunities and vulnerabilities are created by increased reliance on IT	24.3	68.4
	Integrating IO and C4 to support National Military and National Security Strategies	20.8	64.2
	Integrating IO and C4 into the theater and strategic campaign development process	21.9	64.4
	IO, IW, C4I concepts in joint operations	22.8	64.7
Theater Strategy and Campaigning	Role of the unified commander in developing theater plans, policies, and strategies	29.2	68.7
	Coordination of U.S. military plans/actions with foreign forces, interagency, and NGOs	26.6	67.3
	How joint and multinational campaigns and operations support national objectives	25.4	65.2
	Combatant commander's perspective of the resources required to support campaign plans	28.0	67.9
	Organization, responsibilities, and capabilities of military forces available to the JFCs	28.1	66.4
Geo-Strategic Context	Current social, cultural, political, economic, military, technological, and historical issues	25.6	69.3
	Roles and influence of international organizations and other nonstate actors	20.2	64.2
	Key military, nonmilitary, and transnational challenges to U.S. national security	23.0	64.8
Instruments of National Power	Fundamental characteristics, capabilities, and limitations of instruments of national power	19.6	62.6
	Employing diplomatic, economic, military, and informational instruments of national power	19.9	61.9
Joint Operation Art	Joint doctrine and the joint operational art	27.0	66.7
	Integration of service, joint, interagency, and multinational capabilities	30.7	68.3
Joint Warfare Fundamentals	Each combatant command's mission, organization and responsibilities	28.2	68.5
	Joint aspects of military operations other than war (MOOTW)	23.0	64.9
	Capabilities of other services' weapons systems	20.6	65.1

Table C.2—Continued

Broad Knowledge Categories	Specific Knowledge Categories	Required	Required or Helpful
Joint Campaigning	JTF organization, including who can form a JTF and how and when a JTF is formed	22.9	63.3
	Characteristics of a joint campaign and the relationships of supporting capabilities	22.6	62.3
Joint Doctrine	Current joint doctrine	30.0	71.1
	Factors influencing joint doctrine	21.6	65.4
	Relationship between service doctrine and joint doctrine	25.5	68.2
Joint and Multinational Forces at the Operational Level of War	Considerations for employing joint and multinational forces at the operational level of war	20.9	62.8
	How theory and principles of war apply at the operational level of war	18.2	60.5
	Relationships among national objectives, military objectives, and conflict termination	19.7	60.8
	Relationships among the strategic, operational, and tactical levels of war	25.8	63.6
Joint Planning and Execution Processes	Relationships between national objectives and means availability	19.5	61.6
	Effect of time, coordination, policy changes, and political developments on the planning process	19.9	61.5
	How national, joint, and service intelligence organizations support JFCs	21.2	61.9
	Integrating battlespace support systems into campaign/theater planning and operations	18.4	60.4
Other Knowledge Types	Inspector General activities, legal/legislative, law enforcement, physical security, or investigations	12.5	60.9
	Special operations, operations other than war, tactical matters (i.e., training exercises, etc.)	22.2	65.5
	Manpower/personnel, training, education, logistics, acquisition, or general administration	29.5	73.8
	Research and development, engineering, scientific matters (includes weather, environment, etc.), CBRNE matters	14.4	60.9
	Medical or health services	7.8	55.0
	Acquisition/Joint Program Management	20.5	65.5

Table C.3
Percentage of Officers Reporting that Specific Types of Knowledge Are Required or Helpful, by JDAL Category

Broad Knowledge Categories	Specific Knowledge Categories	JDAL Billets	Non-JDAL Billets in External Organizations	Service-Nominated Billets
National Military Capabilities, Organization, and Command Structure	Roles, relationships, and functions of the NCA, JCS, COCOMs, NSC, JFC, CJCS, or the interagency process	95.8	87.3	75.4
	Force structure requirements and resultant capabilities and limitations of U.S. military forces	89.7	79.4	76.5
	How the U.S. military plans, executes, and trains for joint, interagency, multinational operations	91.6	81.8	77.5
	Service-unique capability, limitation, doctrine, and command structure integration	90.0	81.0	79.6
National Military Strategy	Art and science of developing, deploying, employing, and sustaining military resources	84.8	73.3	72.0
	Capabilities and limitations of the U.S. force structure and their effect on joint military strategy	84.4	71.4	67.8
	Concepts of the strategic decisionmaking and defense planning processes	84.9	71.7	64.9
	Resource needs, both national and international, for national defense	83.0	71.2	65.1
	Key considerations that shape the development of national military strategy	82.0	67.1	60.6
	Current National Military Strategy and other examples of U.S. and foreign military strategies	84.6	69.3	62.6
	DoD long-term and immediate process for strategic planning and assessment	81.6	68.7	61.0
National Security Strategy	The national security policy process, to include the integration of national instruments of power	81.5	67.3	58.7
	The impact of defense acquisition and its implications for enhancing joint military capabilities	78.7	69.4	63.1
	Relationships between the military, Congress, NSC, DoD agencies, and the public	84.6	74.8	64.8
	Developing, applying, and coordinating the instruments of national power	77.9	63.4	56.4
	How national policy is turned into executable military strategies	78.5	64.2	56.7
	Capabilities and vulnerabilities of U.S. industry and infrastructure in a global market	73.3	63.6	56.3

Table C.3—Continued

Broad Knowledge Categories	Specific Knowledge Categories	JDAL Billets	Non-JDAL Billets in External Organizations	Service-Nominated Billets
National Security Strategy (cont.)	The national security technological environment for current and future competitive advantage	72.2	62.5	55.9
National Security Policy Process	Origins, responsibilities, organization, and modus operandi of the NSC system	75.3	61.5	51.3
	How major governmental and NGOs influence and implement national security policies	74.5	62.2	52.8
	How the U.S. government prioritizes among issues for developing national-level strategies	73.7	60.9	51.9
National Planning Systems and Processes	National security decisionmaking and the policy formulation process	76.6	62.6	53.0
	Responsibilities and relationships of the interagency and joint community	82.4	70.7	58.3
	DoD processes by which national ends, ways, and means are reconciled, integrated, and applied	74.7	61.9	53.0
	How time, coordination, policy, politics, doctrine, and national power affect the planning process	75.8	64.0	56.0
Command, Control, Communications, Computers, Intelligence, Surveillance, and Reconnaissance (C4ISR)	How C4ISR systems apply at the tactical and operational levels of war	79.7	72.0	71.2
	How IO is incorporated into both deliberate and crisis action-planning processes	77.2	66.7	64.9
	How opportunities and vulnerabilities are created by increased reliance on IT	74.8	68.3	64.4
	Integrating IO and C4 to support National Military and National Security Strategies	73.0	64.4	58.7
	Integrating IO and C4 into the theater and strategic campaign development process	73.3	64.1	59.1
	IO, IW, C4I concepts in joint operations	74.2	64.6	58.9
Theater Strategy and Campaigning	Role of the unified commander in developing theater plans, policies, and strategies	83.1	68.1	60.2
	Coordination of U.S. military plans/actions with foreign forces, interagency, and NGOs	80.1	67.5	59.5
	How joint and multinational campaigns and operations support national objectives	77.8	65.5	57.5

Table C.3—Continued

Broad Knowledge Categories	Specific Knowledge Categories	JDAL Billets	Non-JDAL Billets in External Organizations	Service-Nominated Billets
Theater Strategy and Campaigning (cont.)	Combatant commander's perspective of the resources required to support campaign plans	80.0	66.9	61.0
	Organization, responsibilities, and capabilities of military forces available to the JFCs	78.7	65.2	59.6
Geo-Strategic Context	Current social, cultural, political, economic, military, technological, and historical issues	81.9	70.3	61.3
	Roles and influence of international organizations and other nonstate actors	77.5	65.3	55.7
	Key military, nonmilitary, and transnational challenges to U.S. national security	78.2	65.7	56.3
Instruments of National Power	Fundamental characteristics, capabilities, and limitations of instruments of national power	76.6	61.8	54.5
	Employing diplomatic, economic, military, and informational instruments of national power	76.7	61.6	53.0
Joint Operation Art	Joint doctrine and the joint operational art	78.2	65.9	60.2
	Integration of service, joint, interagency, and multinational capabilities	80.3	68.4	60.8
Joint Warfare Fundamentals	Each combatant command's mission, organization, and responsibilities	81.4	69.1	60.5
	Joint aspects of military operations other than war (MOOTW)	75.3	65.2	58.4
	Capabilities of other services' weapons systems	74.7	63.0	60.3
Joint Campaigning	JTF organization, including who can form a JTF and how and when a JTF is formed	74.7	63.9	56.2
	Characteristics of a joint campaign and the relationships of supporting capabilities	73.0	63.0	55.5
Joint Doctrine	Current joint doctrine	82.7	70.4	64.1
	Factors influencing joint doctrine	77.3	65.4	58.1
	Relationship between service doctrine and joint doctrine	78.7	68.1	62.0
Joint and Multinational Forces at the Operational Level of War	Considerations for employing joint and multinational forces at the operational level of war	74	61.5	56.5
	How theory and principles of war apply at the operational level of war	71.3	58.8	54.7

Table C.3—Continued

Broad Knowledge Categories	Specific Knowledge Categories	JDAL Billets	Non-JDAL Billets in External Organizations	Service-Nominated Billets
Joint and Multinational Forces at the Operational Level of War (cont.)	Relationships among national objectives, military objectives, and conflict termination	72.6	59.6	54.2
	Relationships among the strategic, operational, and tactical levels of war	74.8	62.1	57.4
Joint Planning and Execution Processes	Relationships between national objectives and means availability	75.1	60.5	53.9
	Effect of time, coordination, policy changes, and political developments on the planning process	74.0	61.0	54.1
	How national, joint, and service intelligence organizations support JFCs	74.6	62.2	54.1
	Integrating battlespace support systems into campaign/theater planning and operations	71.1	59.3	54.4
Other Knowledge Types	Inspector General activities, legal/legislative, law enforcement, physical security, or investigations	67.5	59.6	57.4
	Special operations, operations other than war, tactical matters (i.e., training exercises, etc.)	72.7	62.3	62.4
	Manpower/personnel, training, education, logistics, acquisition, or general administration	76.0	72.2	73.2
	Research and development, engineering, scientific matters (includes weather, environment, etc.), CBRNE matters	67.3	60.4	57.3
	Medical or health services	62.2	53.9	51.2
	Acquisition/Joint Program Management	69.8	67.7	62.0

Table C.4.
Percentage of Officers Reporting that Specific Types of Knowledge Are Required or Helpful, by Billet Paygrade

Broad Knowledge Categories	Specific Knowledge Categories	Junior Officers	Mid-level Officers	Flag Officers
National Military Capabilities, Organization and Command Structure	Roles, Relationships, and Functions of the NCA, JCS, COCOMs, NSC, JFC, CJCS, or the interagency process	70.1	88.0	94.9
	Force structure requirements and resultant capabilities and limitations of U.S. military forces	69.9	83.9	95.2
	How the U.S. military plans, executes, and trains for joint, interagency, multinational operations	72.5	85.3	95.8
	Service-unique capability, limitation, doctrine, and command structure integration	72.4	85.8	97.1
National Military Strategy	Art and science of developing, deploying, employing, and sustaining military resources	63.7	79.2	93.8
	Capabilities and limitations of the U.S. force structure and their effect on joint military strategy	60.1	77.0	91.1
	Concepts of the strategic decisionmaking and defense planning processes	58.2	76.1	89.9
	Resource needs, both national and international, for national defense	58.0	75.1	92.2
	Key considerations that shape the development of national military strategy	54.6	71.8	88.7
	Current National Military Strategy and other examples of U.S. and foreign military strategies	56.2	74.2	90.2
	DoD long-term and immediate process for strategic planning and assessment	55.2	72.3	87.0
National Security Strategy	The national security policy process, to include the integration of national instruments of power	53.0	71.0	87.9
	The impact of defense acquisition and its implications for enhancing joint military capabilities	55.3	72.6	89.3
	Relationships between the military, Congress, NSC, DoD agencies, and the public	57.8	76.7	93.7
	Developing, applying, and coordinating the instruments of national power	50.7	67.7	85.9
	How national policy is turned into executable military strategies	51.4	68.2	86.1
	Capabilities and vulnerabilities of U.S. industry and infrastructure in a global market	50.2	66.0	84.9
	The national security technological environment for current and future competitive advantage	50.3	65.1	83.2
National Security Policy Process	Origins, responsibilities, organization, and modus operandi of the NSC system	47.8	63.9	78.9

Table C.4—Continued

Broad Knowledge Categories	Specific Knowledge Categories	Junior Officers	Mid-level Officers	Flag Officers
National Security Policy Process (cont.)	How major governmental and NGOs influence and implement national security policies	48.6	64.5	79.4
	How the U.S. government prioritizes among issues for developing national-level strategies	47.9	63.6	77.5
National Planning Systems and Processes	National security decisionmaking and the policy formulation process	47.9	65.7	83.0
	Responsibilities and relationships of the interagency and joint community	53.4	72.1	87.0
	DoD processes by which national ends, ways, and means are reconciled, integrated, and applied	47.3	65.0	81.0
	How time, coordination, policy, politics, doctrine and national power affect the planning process	50.2	67.1	82.3
Command, Control, Communications, Computers, Intelligence, Surveillance, and Reconnaissance (C4ISR)	How C4ISR systems apply at the tactical and operational levels of war	66.9	75.6	85.2
	How IO is incorporated into both deliberate and crisis action-planning processes	61.0	70.9	82.2
	How opportunities and vulnerabilities are created by increased reliance on IT	60.8	70.1	82.8
	Integrating IO and C4 to support National Military and National Security Strategies	55.6	66.3	78.1
	Integrating IO and C4 into the theater and strategic campaign development process	55.5	66.6	77.8
	IO, IW, C4I concepts in joint operations	55.9	66.9	77.9
Theater Strategy and Campaigning	Role of the unified commander in developing theater plans, policies, and strategies	54.2	72.6	85.0
	Coordination of U.S. military plans/actions with foreign forces, interagency, and NGOs	55.2	70.6	83.0
	How joint and multinational campaigns and operations support national objectives	53.2	68.3	82.3
	Combatant commander's perspective of the resources required to support campaign plans	54.6	71.3	87.0
	Organization, responsibilities, and capabilities of military forces available to the JFCs	52.8	70.0	84.7
Geo-Strategic Context	Current social, cultural, political, economic, military, technological, and historical issues	57.8	72.3	85.9
	Roles and influence of international organizations and other nonstate actors	52.2	67.3	81.1

Table C.4—Continued

Broad Knowledge Categories	Specific Knowledge Categories	Junior Officers	Mid-level Officers	Flag Officers
Geo-Strategic Context (cont.)	Key military, nonmilitary, and transnational challenges to U.S. national security	52.9	67.9	82.6
Instruments of National Power	Fundamental characteristics, capabilities, and limitations of instruments of national power	49.4	66.1	81.3
	Employing diplomatic, economic, military, and informational instruments of national power	48.2	65.5	81.0
Joint Operation Art	Joint doctrine and the joint operational art	53.8	70.1	83.6
	Integration of service, joint, interagency, and multinational capabilities	54.6	71.9	83.6
Joint Warfare Fundamentals	Each combatant command's mission, organization, and responsibilities	55.4	72.1	84.1
	Joint aspects of military operations other than war (MOOTW)	53.2	67.9	83.3
	Capabilities of other services' weapons systems	54.3	67.9	81
Joint Campaigning	JTF organization, including who can form a JTF and how and when a JTF is formed	51.0	66.5	82.2
	Characteristics of a joint campaign and the relationships of supporting capabilities	50.1	65.4	81.1
Joint Doctrine	Current joint doctrine	57.2	74.9	86.6
	Factors influencing joint doctrine	53.0	68.6	82.1
	Relationship between service doctrine and joint doctrine	55.2	71.6	85.4
Joint and Multinational Forces at the Operational Level of War	Considerations for employing joint and multinational forces at the operational level of war	50.2	66.0	81.4
	How theory and principles of war apply at the operational level of war	49.1	63.4	79.7
	Relationships among national objectives, military objectives, and conflict termination	48.9	63.8	79.9
	Relationships among the strategic, operational, and tactical levels of war	51.4	66.7	83.4
Joint Planning and Execution Processes	Relationships between national objectives and means availability	49.0	65.0	78.9
	Effect of time, coordination, policy changes, and political developments on the planning process	48.9	64.8	79.1
	How national, joint, and service intelligence organizations support JFCs	49.7	65	80.2
	Integrating battlespace support systems into campaign/theater planning and operations	48.9	63.2	79.1

Table C.4—Continued

Broad Knowledge Categories	Specific Knowledge Categories	Junior Officers	Mid-level Officers	Flag Officers
Other Knowledge Types	Inspector General activities, legal/legislative, law enforcement, physical security, or investigations	52.1	62.8	83.3
	Special operations, operations other than war, tactical matters (i.e., training exercises, etc.)	58.0	67.1	82.0
	Manpower/personnel, training, education, logistics, acquisition, or general administration	66.5	75.3	93.0
	Research and development, engineering, scientific matters (includes weather, environment, etc.), CBRNE matters	50.3	63.4	82.3
	Medical or health services	46.1	56.9	77.2
	Acquisition/Joint Program Management	55.5	68.0	82.7

Table C.5
Percentage of Officers Reporting that Their Assignment Will Lead to "Proficiency" or "Familiarity" with Specific Types of Knowledge

Broad Knowledge Categories	Specific Knowledge Categories	Reporting that Assignment Will Lead to "Proficiency"	Reporting that Assignment Will Lead to "Proficiency or Familiarity"
National Military Capabilities, Organization, and Command Structure	Roles, Relationships, and Functions of the NCA, JCS, COCOMs, NSC, JFC, CJCS, or the interagency process	42.2	81.6
	Force structure requirements and resultant capabilities and limitations of U.S. military forces	35.9	78.3
	How the U.S. military plans, executes, and trains for joint, interagency, multinational operations	37.9	79.6
	Service-unique capability, limitation, doctrine, and command structure integration	36.1	80.2
National Military Strategy	Art and science of developing, deploying, employing, and sustaining military resources	28.2	73.2
	Capabilities and limitations of the U.S. force structure and their effect on joint military strategy	25.9	70.8
	Concepts of the strategic decisionmaking and defense planning processes	26.6	69.7
	Resource needs, both national and international, for national defense	22.0	69.5
	Key considerations that shape the development of national military strategy	20.1	66.0
	Current National Military Strategy and other examples of U.S. and foreign military strategies	24.0	67.9
	DoD long-term and immediate process for strategic planning and assessment	20.1	66.4
National Security Strategy	The national security policy process, to include the integration of national instruments of power	20.0	64.3
	The impact of defense acquisition and its implications for enhancing joint military capabilities	21.2	66.9
	Relationships between the military, Congress, NSC, DoD agencies, and the public	27.5	70.6
	Developing, applying, and coordinating the instruments of national power	18.0	62.1
	How national policy is turned into executable military strategies	19.0	62.5
	Capabilities and vulnerabilities of U.S. industry and infrastructure in a global market	13.6	60.9

Table C.5—Continued

Broad Knowledge Categories	Specific Knowledge Categories	Reporting that Assignment Will Lead to "Proficiency"	Reporting that Assignment Will Lead to "Proficiency or Familiarity"
National Security Strategy (cont.)	The national security technological environment for current and future competitive advantage	14.1	60.4
National Security Policy Process	Origins, responsibilities, organization, and modus operandi of the NSC system	11.5	58.1
	How major governmental and NGOs influence and implement national security policies	12.1	59.1
	How the U.S. government prioritizes among issues for developing national-level strategies	11.7	58.0
	National security decisionmaking and the policy formulation process	14.1	59.3
National Planning Systems and Processes	Responsibilities and relationships of the interagency and joint community	24.7	65.7
	DoD processes by which national ends, ways, and means are reconciled, integrated, and applied	15.0	58.9
	How time, coordination, policy, politics, doctrine and national power affect the planning process	18.0	61.5
Command, Control, Communications, Computers, Intelligence, Surveillance, and Reconnaissance (C4ISR)	How C4ISR systems apply at the tactical and operational levels of war	30.0	71.2
	How IO is incorporated into both deliberate and crisis action-planning processes	23.1	66.4
	How opportunities and vulnerabilities are created by increased reliance on IT	22.3	66.0
	Integrating IO and C4 to support National Military and National Security Strategies	18.5	61.8
	Integrating IO and C4 into the theater and strategic campaign development process	19.5	62.2
	IO, IW, C4I concepts in joint operations	20.1	62.4
Theater Strategy and Campaigning	Role of the unified commander in developing theater plans, policies, and strategies	26.1	65.7
	Coordination of U.S. military plans/actions with foreign forces, interagency, and NGOs	24.0	64.6
	How joint and multinational campaigns and operations support national objectives	23.7	62.9

Table C.5—Continued

Broad Knowledge Categories	Specific Knowledge Categories	Reporting that Assignment Will Lead to "Proficiency"	Reporting that Assignment Will Lead to "Proficiency or Familiarity"
Theater Strategy and Campaigning (cont.)	Combatant commander's perspective of the resources required to support campaign plans	25.0	65.1
	Organization, responsibilities, and capabilities of military forces available to the JFCs	25.3	64.1
Geo-Strategic Context	Current social, cultural, political, economic, military, technological, and historical issues	24.4	66.7
	Roles and influence of international organizations and other nonstate actors	19.0	61.9
	Key military, nonmilitary, and transnational challenges to U.S. national security	21.3	62.6
Instruments of National Power	Fundamental characteristics, capabilities, and limitations of instruments of national power	18.7	59.9
	Employing diplomatic, economic, military, and informational instruments of national power	19.1	59.7
Joint Operation Art	Joint doctrine and the joint operational art	24.2	64.0
	Integration of service, joint, interagency, and multinational capabilities	27.4	65.8
Joint Warfare Fundamentals	Each combatant command's mission, organization, and responsibilities	26.1	65.7
	Joint aspects of military operations other than war (MOOTW)	21.1	62.4
	Capabilities of other services' weapons systems	19.0	62.8
Joint Campaigning	JTF organization, including who can form a JTF and how and when a JTF is formed	21.4	60.8
	Characteristics of a joint campaign and the relationships of supporting capabilities	20.9	59.8
Joint Doctrine	Current joint doctrine	24.9	67.3
	Factors influencing joint doctrine	19.5	62.2
	Relationship between service doctrine and joint doctrine	21.8	65.0
Joint and Multinational Forces at the Operational Level of War	Considerations for employing joint and multinational forces at the operational level of war	18.9	59.6
	How theory and principles of war apply at the operational level of war	17.8	57.8

Table C.5—Continued

Broad Knowledge Categories	Specific Knowledge Categories	Reporting that Assignment Will Lead to "Proficiency"	Reporting that Assignment Will Lead to "Proficiency or Familiarity"
Joint and Multinational Forces at the Operational Level of War (cont.)	Relationships among national objectives, military objectives, and conflict termination	18.6	58.1
	Relationships among the strategic, operational, and tactical levels of war	24.6	60.9
Joint Planning and Execution Processes	Relationships between national objectives and means availability	17.4	58.1
	Effect of time, coordination, policy changes, and political developments on the planning process	18.1	58.5
	How national, joint, and service intelligence organizations support JFCs	18.9	59.0
	Integrating battlespace support systems into campaign/theater planning and operations	16.2	57.6
Other Knowledge Types	Inspector General activities, legal/legislative, law enforcement, physical security, or investigations	11.9	58.1
	Special operations, operations other than war, tactical matters (i.e., training exercises, etc.)	20.3	62.6
	Manpower/personnel, training, education, logistics, acquisition, or general administration	27.9	70.8
	Research and development, engineering, scientific matters (includes weather, environment, etc.), CBRNE matters	13.7	58.9
	Medical or health services	7.5	53.4
	Acquisition/Joint Program Management	19.8	63.8

Table C.6
Percentage of Officers Reporting that Their Assignment Will Lead to "Proficiency" or "Familiarity" with Specific Types of Knowledge, by JDAL Category

Broad Knowledge Categories	Specific Knowledge Categories	JDAL Billets	Non-JDAL Billets in External Organizations	Service-Nominated Billets
National Military Capabilities, Organization and Command Structure	Roles, Relationships, and Functions of the NCA, JCS, COCOMs, NSC, JFC, CJCS, or the interagency process	93.8	71.1	85.2
	Force structure requirements and resultant capabilities and limitations of U.S. military forces	87.5	72.4	76.9
	How the U.S. military plans, executes, and trains for joint, interagency, multinational operations	89.3	71.2	78.8
	Service-unique capability, limitation, doctrine, and command structure integration	88.0	77.3	78.2
National Military Strategy	Art and science of developing, deploying, employing, and sustaining military resources	82.3	70.0	70.5
	Capabilities and limitations of the U.S. force structure and their effect on joint military strategy	82.3	68.4	68.8
	Concepts of the strategic decisionmaking and defense planning processes	82.7	65.6	69.0
	Resource needs, both national and international, for national defense	81.0	69.0	68.7
	Key considerations that shape the development of national military strategy	80.0	65.1	64.9
	Current National Military Strategy and other examples of U.S. and foreign military strategies	82.2	64.4	66.7
	DoD long-term and immediate process for strategic planning and assessment	79.6	66.2	66.0
National Security Strategy	The national security policy process, to include the integration of national instruments of power	78.8	63.3	63.9
	The impact of defense acquisition and its implications for enhancing joint military capabilities	76.8	72.2	66.8
	Relationships between the military, Congress, NSC, DoD agencies, and the public	82.2	71.7	72.6
	Developing, applying, and coordinating the instruments of national power	75.9	61.8	61.1
	How national policy is turned into executable military strategies	76.1	62.0	62.1

Table C.6—Continued

Broad Knowledge Categories	Specific Knowledge Categories	JDAL Billets	Non-JDAL Billets in External Organizations	Service-Nominated Billets
National Security Strategy (cont.)	Capabilities and vulnerabilities of U.S. industry and infrastructure in a global market	71.2	67.4	61.7
	The national security technological environment for current and future competitive advantage	70.4	66.7	60.6
National Security Policy Process	Origins, responsibilities, organization, and modus operandi of the NSC system	72.6	64.5	58.3
	How major governmental and NGOs influence and implement national security policies	71.9	67.0	59.8
	How the U.S. government prioritizes among issues for developing national-level strategies	71.3	64.3	58.4
National Planning Systems and Processes	National security decisionmaking and the policy formulation process	73.7	63.4	59.5
	Responsibilities and relationships of the interagency and joint community	80.1	67.0	68.2
	DoD processes by which national ends, ways, and means are reconciled, integrated, and applied	72.1	61.8	59.1
	How time, coordination, policy, politics, doctrine, and national power affect the planning process	74.0	61.8	61.8
Command, Control, Communications, Computers, Intelligence, Surveillance, and Reconnaissance (C4ISR)	How C4ISR systems apply at the tactical and operational levels of war	77.4	72.7	68.5
	How IO is incorporated into both deliberate and crisis action-planning processes	74.5	69.2	64.2
	How opportunities and vulnerabilities are created by increased reliance on IT	73.0	70.6	65.7
	Integrating IO and C4 to support National Military and National Security Strategies	71.0	67.9	61.6
	Integrating IO and C4 into the theater and strategic campaign development process	71.0	67.0	61.6
	IO, IW, C4I concepts in joint operations	72.3	66.5	62.4
Theater Strategy and Campaigning	Role of the unified commander in developing theater plans, policies, and strategies	80.1	62.4	65.0

Table C.6—Continued

Broad Knowledge Categories	Specific Knowledge Categories	JDAL Billets	Non-JDAL Billets in External Organizations	Service-Nominated Billets
Theater Strategy and Campaigning (cont.)	Coordination of U.S. military plans/actions with foreign forces, interagency, and NGOs	77.6	64.4	64.4
	How joint and multinational campaigns and operations support national objectives	75.7	62.3	63.0
	Combatant commander's perspective of the resources required to support campaign plans	77.6	63.3	64.0
	Organization, responsibilities, and capabilities of military forces available to the JFCs	76.8	62.8	62.9
Geo-Strategic Context	Current social, cultural, political, economic, military, technological, and historical issues	79.5	67.8	67.5
	Roles and influence of international organizations and other nonstate actors	75.3	65.3	62.7
	Key military, nonmilitary, and transnational challenges to U.S . national security	76.1	63.5	63.4
Instruments of National Power	Fundamental characteristics, capabilities, and limitations of instruments of national power	74.0	60.6	59.0
	Employing diplomatic, economic, military, and informational instruments of national power	74.4	60.7	59.3
Joint Operation Art	Joint doctrine and the joint operational art	75.8	62.8	62.6
	Integration of service, joint, interagency, and multinational capabilities	77.9	65.1	66.0
Joint Warfare Fundamentals	Each combatant command's mission, organization, and responsibilities	78.7	63.2	66.0
	Joint aspects of military operations other than war (MOOTW)	73.2	64.6	62.6
	Capabilities of other services' weapons systems	72.6	64.3	60.7
Joint Campaigning	JTF organization, including who can form a JTF and how and when a JTF is formed	72.4	60.5	60.5
	Characteristics of a joint campaign and the relationships of supporting capabilities	70.6	59.8	60.0
Joint Doctrine	Current joint doctrine	79.5	67.3	66.6
	Factors influencing joint doctrine	74.2	64.7	61.8

Table C.6—Continued

Broad Knowledge Categories	Specific Knowledge Categories	JDAL Billets	Non-JDAL Billets in External Organizations	Service-Nominated Billets
Joint Doctrine (cont.)	Relationship between service doctrine and joint doctrine	75.5	66.9	64.6
Joint and Multinational Forces at the Operational Level of War	Considerations for employing joint and multinational forces at the operational level of war	70.8	59.6	58.7
	How theory and principles of war apply at the operational level of war	68.4	57.9	56.1
	Relationships among national objectives, military objectives, and conflict termination	70.2	58.5	56.5
	Relationships among the strategic, operational, and tactical levels of war	72.4	58.3	59.4
Joint Planning and Execution Processes	Relationships between national objectives and means availability	71.3	59.7	57.1
	Effect of time, coordination, policy changes, and political developments on the planning process	71.1	59.6	58.0
	How national, joint, and service intelligence organizations support JFCs	71.7	62.7	59.5
	Integrating battlespace support systems into campaign/theater planning and operations	68. 4	60.9	56.1
Other Knowledge Types	Inspector General activities, legal/legislative, law enforcement, physical security, or investigations	65.0	66.5	56.9
	Special operations, operations other than war, tactical matters (i.e., training exercises, etc.)	70.4	61.0	59.2
	Manpower/personnel, training, education, logistics, acquisition, or general administration	73.6	75.5	68.9
	Research and development, engineering, scientific matters (includes weather, environment, etc.), CBRNE matters	65.6	67.2	58.2
	Medical or health services	60.9	63.5	52.3
	Acquisition/Joint Program Management	68.2	73.3	65.8

Table C.7
Percentage of Officers Reporting that Their Assignment Will Lead to "Proficiency" or "Familiarity" with Specific Types of Knowledge, by Billet Paygrade

Broad Knowledge Categories	Specific Knowledge Categories	Junior Officer	Mid-Level Officer	Flag Officer
National Military Capabilities, Organization, and Command Structure	Roles, relationships, and functions of the NCA, JCS, COCOMs, NSC, JFC, CJCS, or the interagency process	66.9	85.9	93.1
	Force structure requirements and resultant capabilities and limitations of U.S. military forces	66.2	81.6	92.7
	How the U.S. military plans, executes, and trains for joint, interagency, multinational operations	68.0	82.7	93.0
	Service-unique capability, limitation, doctrine, and command structure integration	68.3	83.4	94.8
National Military Strategy	Art and science of developing, deploying, employing, and sustaining military resources	60.5	76.4	91.6
	Capabilities and limitations of the U.S. force structure and their effect on joint military strategy	57.0	74.6	87.5
	Concepts of the strategic decisionmaking and defense planning processes	55.3	73.5	86.9
	Resource needs, both national and international, for national defense	55.3	73.0	89.8
	Key considerations that shape the development of national military strategy	51.8	69.8	85.8
	Current National Military Strategy and other examples of U.S. and foreign military strategies	53.4	71.8	87.6
	DoD long-term and immediate process for strategic planning and assessment	52.4	70.0	85.8
National Security Strategy	The national security policy process, to include the integration of national instruments of power	49.7	68.1	85.1
	The impact of defense acquisition and its implications for enhancing joint military capabilities	53	70.5	86.8
	Relationships between the military, Congress, NSC, DoD agencies, and the public	55.4	74.6	91.9
	Developing, applying, and coordinating the instruments of national power	48.2	65.7	82.6
	How national policy is turned into executable military strategies	48.9	66	83.8
	Capabilities and vulnerabilities of U.S. industry and infrastructure in a global market	48.2	64.1	82.4
	The national security technological environment for current and future competitive advantage	48.3	63.3	81.2
National Security Policy Process	Origins, responsibilities, organization, and modus operandi of the NSC system	45	61.4	77

Table C.7—Continued

Broad Knowledge Categories	Specific Knowledge Categories	Junior Officer	Mid-Level Officer	Flag Officer
National Security Policy Process (cont.)	How major governmental and NGOs influence and implement national security policies	46.2	62.4	77.8
	How the U.S. government prioritizes among issues for developing national-level strategies	45.4	61.2	75.9
National Planning Systems and Processes	National security decisionmaking and the policy formulation process	45.2	62.9	80.2
	Responsibilities and relationships of the interagency and joint community	50.9	69.6	83.8
	DoD processes by which national ends, ways, and means are reconciled, integrated, and applied	45.1	62.5	77.8
	How time, coordination, policy, politics, doctrine, and national power affect the planning process	47.7	65.1	79.6
Command, Control, Communications, Computers, Intelligence, Surveillance, and Reconnaissance (C4ISR)	How C4ISR systems apply at the tactical and operational levels of war	63.7	73.1	83.7
	How IO is incorporated into both deliberate and crisis action-planning processes	57.7	68.5	80.5
	How opportunities and vulnerabilities are created by increased reliance on IT	57.6	67.9	81
	Integrating IO and C4 to support National Military and National Security Strategies	52.8	64.1	76.5
	Integrating IO and C4 into the theater and strategic campaign development process	52.9	64.5	76.2
	IO, IW, C4I concepts in joint operations	52.9	64.9	75.9
Theater Strategy and Campaigning	Role of the unified commander in developing theater plans, policies, and strategies	51.1	69.7	80.7
	Coordination of U.S. military plans/actions with foreign forces, interagency, and NGOs	52	68	78.9
	How joint and multinational campaigns and operations support national objectives	50.4	66.1	78.6
	Combatant commander's perspective of the resources required to support campaign plans	51.5	68.8	83.6
	Organization, responsibilities, and capabilities of military forces available to the JFCs	50.3	67.7	81.1
Geo-Strategic Context	Current social, cultural, political, economic, military, technological, and historical issues	54.4	69.9	83.6
	Roles and influence of international organizations and other nonstate actors	49.4	65.2	79.4

Table C.7—Continued

Broad Knowledge Categories	Specific Knowledge Categories	Junior Officer	Mid-Level Officer	Flag Officer
Geo-Strategic Context	Key military, nonmilitary, and transnational challenges to U.S. national security	49.6	65.8	80.9
Instruments of National Power	Fundamental characteristics, capabilities, and limitations of instruments of national power	46.5	63.4	78.3
	Employing diplomatic, economic, military and informational instruments of national power	45.7	63.4	78.1
Joint Operation Art	Joint doctrine and the joint operational art	50.4	67.6	80.6
	Integration of service, joint, interagency, and multinational capabilities	51.6	69.7	81.2
Joint Warfare Fundamentals	Each Combatant Command's mission, organization, and responsibilities	52	69.3	81.7
	Joint aspects of military operations other than war (MOOTW)	49.7	65.7	80.4
	Capabilities of other services' weapons systems	51.2	65.8	79.2
Joint Campaigning	JTF organization, including who can form a JTF and how and when a JTF is formed	47.9	64.1	79.8
	Characteristics of a joint campaign and the relationships of supporting capabilities	47	63.1	78.4
Joint Doctrine	Current joint doctrine	53	71.3	83.3
	Factors influencing joint doctrine	49.2	65.7	79.4
	Relationship between service doctrine and joint doctrine	51.2	68.6	83.7
Joint and Multinational Forces at the Operational Level of War	Considerations for employing joint and multinational forces at the operational level of war	46.5	63	78.3
	How theory and principles of war apply at the operational level of war	45.6	61	76.3
	Relationships among national objectives, military objectives, and conflict termination	45.2	61.5	76.3
	Relationships among the strategic, operational, and tactical levels of war	47.7	64.2	80.7
Joint Planning and Execution Processes	Relationships between national objectives and means availability	45.1	61.5	75.5
	Effect of time, coordination, policy changes, and political developments on the planning process	45.7	61.9	75.8
	How national, joint, and service intelligence organizations support JFCs	46.7	62.3	75.8
	Integrating battlespace support systems into campaign/theater planning and operations	45.7	60.6	75.9

Table C.7—Continued

Broad Knowledge Categories	Specific Knowledge Categories	Junior Officer	Mid-Level Officer	Flag Officer
Other Knowledge Types	Inspector General activities, legal/legislative, law enforcement, physical security or investigations	48.8	60.1	80.8
	Special operations, operations other than war, tactical matters (i.e., training exercises, etc.)	54.2	64.7	79.1
	Manpower/personnel, training, education, logistics, acquisition, or general administration	62.4	72.5	90.9
	Research and development, engineering, scientific matters (includes weather, environment, etc.), CBRNE matters	47.9	61.3	80.1
	Medical or health services	43.6	55.6	75.8
	Acquisition/Joint Program Management	53.5	66.4	80.4

Bibliography

Booz Allen Hamilton, *Independent Study of Joint Officer Management and Joint Professional Military Education,* Booz Allen Hamilton: McLean, Va., 2003.

Chairman of the Joint Chiefs of Staff (CJCS) Instruction 1331.01B, *Manpower and Personnel Actions Involving General and Flag Officers,* August 29, 2003.

CJCS Instruction 1800.01B, *Officer Professional Military Education Policy,* August 30, 2004.

DoD Instruction (DODI) 1300.20, *DoD Joint Officer Management Program Procedures,* December 20, 1996.

Goldwater-Nichols Department of Defense Reorganization Act of 1986, Public Law 99-433, October 1, 1986.

Harrell, Margaret C., John F. Schank, Harry J. Thie, Clifford M. Graff II, and Paul Steinberg, *How Many Can Be Joint? Supporting Joint Duty Assignments,* Santa Monica, Calif.: RAND Corporation, MR-593-JS, 1996.

Joint Chiefs of Staff, *Joint Vision 2010,* Washington, D.C.: Office of the Chairman of the Joint Chiefs of Staff, 1995.

———, *Joint Vision 2020,* Washington, D.C.: Office of the Chairman of the Joint Chiefs of Staff, June 2000.

———, *An Evolving Joint Perspective: U.S. Joint Warfare and Crisis Resolution in the 21st Century,* Washington, D.C.: Joint Vision and Transformation Division, 2003.

Joint Staff Memorandum, *Cumulative Joint Duty Credit for Combined Joint Task Force Headquarters Assignments in Approved Operations,* J-1A 00165-04, July 12, 2004.

Locher, James R. III, "Has It Worked? The Goldwater-Nichols Reorganization Act," *Naval War College Review,* Vol. 54, No. 4, Autumn 2001.

MacGregor, Douglas A., "The Joint Force: A Decade, No Progress," *Joint Force Quarterly,* Vol. 27, Winter 2000–2001.

Rumsfeld, Donald H., Secretary of Defense, *Annual Report to the President and the Congress,* 2004 (available at http://www.defenselink.mil/execsec/adr2004/index.html as of January 2006).

Schank, John F., Harry J. Thie, and Margaret C. Harrell, *Identifying and Supporting Joint Duty Assignments: Executive Summary,* Santa Monica, Calif.: RAND, MR-622-JS, 1996.

Schank, John F., Harry J. Thie, Jennifer Kawata, Margaret C. Harrell, Clifford M. Graf II, and Paul Steinberg, *Who Is Joint? Reevaluating the Joint Duty Assignment List,* Santa Monica, Calif.: RAND, MR-574-JS, 1996.

Scully, Megan, "Joint But Not Yet Seamless: Military Exercise Exposes Cultural Gaps in U.S. Inter-Service Operations," *Defense News,* February 2, 2004.

Thie, Harry J., Margaret C. Harrell, and Robert M. Emmerichs, *Interagency and International Assignments and Officer Career Management,* Santa Monica, Calif.: RAND Corporation, MR-1116-OSD, 1999.

Thie, Harry J., Margaret C. Harrell, Roland J. Yardley, Marian Oshiro, Holly Ann Potter, Peter Schirmer, and Nelson Lim, *Framing a Strategic Approach for Joint Officer Management,* Santa Monica, Calif.: RAND Corporation, MG-306-OSD, 2005.

U.S. Code, Title 10—Armed Forces, Subtitle A—General Military Law, Part II—Personnel, Chapter 38—Joint Office Management, Section 668—Definitions, (available at http://www.access.gpo.gov/uscode/title10/subtitlea_partii_chapter38.html as of January 2006).

U.S. Department of Defense, Office of the Inspector General, *Audit of the Joint Operation Planning and Execution System,* Report 94-160, Washington, D.C., June 30, 1994.

U.S. General Accounting Office (now the Government Accountability Office), *Joint Officer Development Has Improved, but a Strategic Approach Is Needed,* GAO: Washington, D.C., GAO-03-238, December 2002.

———, *Human Capital: A Guide for Assessing Strategic Training and Development Efforts in the Federal Government,* Washington, D.C.: GAO, GAO-03-893G, 2003a.

———, "A Strategic Approach Is Needed to Improve Joint Officer Development," statement of Derek B. Smith, Director, Defense Capabilities and Management, Testimony before the Subcommittee on Total Force, Committee on Armed Services, House of Representatives, Wednesday, March 19, 2003b.